CONTENTS

Improved *Mycobacterium avium* Complex Detection Methods

Awwa Research Foundation

About the Awwa Research Foundation

The Awwa Research Foundation (AwwaRF) is a member-supported, international, nonprofit organization that sponsors research to enable water utilities, public health agencies, and other professionals to provide safe and affordable drinking water to consumers.

The Foundation's mission is to advance the science of water to improve the quality of life. To achieve this mission, the Foundation sponsors studies on all aspects of drinking water, including supply and resources, treatment, monitoring and analysis, distribution, management, and health effects. Funding for research is provided primarily by subscription payments from approximately 1,000 utilities, consulting firms, and manufacturers in North America and abroad. Additional funding comes from collaborative partnerships with other national and international organizations, allowing for resources to be leveraged, expertise to be shared, and broad-based knowledge to be developed and disseminated. Government funding serves as a third source of research dollars.

From its headquarters in Denver, Colorado, the Foundation's staff directs and supports the efforts of more than 800 volunteers who serve on the board of trustees and various committees. These volunteers represent many facets of the water industry, and contribute their expertise to select and monitor research studies that benefit the entire drinking water community.

The results of research are disseminated through a number of channels, including reports, the Web site, conferences, and periodicals.

For subscribers, the Foundation serves as a cooperative program in which water suppliers unite to pool their resources. By applying Foundation research findings, these water suppliers can save substantial costs and stay on the leading edge of drinking water science and technology. Since its inception, AwwaRF has supplied the water community with more than $300 million in applied research.

More information about the Foundation and how to become a subscriber is available on the Web at **www.awwarf.org.**

Improved *Mycobacterium avium* Complex Detection Methods

Prepared by:
Michael T. Collins and **Alice S. Yuroff**
School of Veterinary Medicine
University of Wisconsin – Madison, 2015 Linden Drive, Madison, Wisconsin 53706

Rebecca M. Hoffman
Wisconsin State Laboratory of Hygiene
University of Wisconsin – Madison, 2601 Agriculture Drive, Madison, Wisconsin 53707

and

Gregory W. Harrington and **Andrew D. Jacque**
Department of Civil and Environmental Engineering
University of Wisconsin – Madison, 1415 Engineering Drive, Madison, Wisconsin 53706

Jointly sponsored by:
Awwa Research Foundation
6666 West Quincy Avenue, Denver, CO 80235-3098

and

U.S. Environmental Protection Agency
Washington, D.C.

Published by:

Awwa Research Foundation

Distributed by:

American Water Works Association

IWA Publishing

DISCLAIMER

This study was jointly funded by the Awwa Research Foundation (AwwaRF) and the U.S. Environmental Protection Agency (USEPA) under Cooperative Agreement No. CR-83110401. AwwaRF and USEPA assume no responsibility for the content of the research study reported in this publication or for the opinions or statements of fact expressed in the report. The mention of trade names for commercial products does not represent or imply the approval or endorsement of AwwaRF or USEAP. This report is presented solely for informational purposes.

ISBN 978-1-84339-627-7

TABLES

FIGURES

FOREWORD

The Awwa Research Foundation is a nonprofit corporation that is dedicated to the implementation of a research effort to help utilities respond to regulatory requirements and traditional high-priority concerns of the industry. The research agenda is developed through a process of consultation with subscribers and drinking water professionals. Under the umbrella of a Strategic Research Plan, the Research Advisory Council prioritizes the suggested projects based upon current and future needs, applicability, and past work; the recommendations are forwarded to the Board of Trustees for final selection. The foundation also sponsors research projects through the unsolicited proposal process; the Collaborative Research, Research Applications, and Tailored Collaboration programs; and various joint research efforts with organizations such as the U.S. Environmental Protection Agency, the U.S. Bureau of Reclamation, and the Association of California Water Agencies.

This publication is a result of one of these sponsored studies, and it is hoped that its findings will be applied in communities throughout the world. The following report serves not only as a means of communicating the results of the water industry's centralized research program but also as a tool to enlist the further support of non member utilities and individuals.

Projects are managed closely from their inception to the final report by the foundation's staff and large cadre of volunteers who willingly contribute their time and expertise. The foundation serves a planning and management function and awards contracts to other institutions such as water utilities, universities, and engineering firms. The funding for this research effort comes primarily from the Subscription Program, through which water utilities subscribe to the research program and make an annual payment proportionate to the volume of water they deliver and consultants and manufacturers subscribe based on their annual billings. The program offers a cost-effective and fair method for funding research in the public interest.

A broad spectrum of water supply issues is addressed by the foundation's research agenda: resources, treatment and operations, distribution and storage, water quality and analysis, toxicology, economics, and management. The ultimate purpose of the coordinated effort is to assist water suppliers to provide the highest possible quality of water economically and reliably. The true benefits are realized when the results are implemented at the utility level. The foundation's trustees are pleased to offer this publication as a contribution toward that end.

David E. Rager
Chair, Board of Trustees
Awwa Research Foundation

Robert C. Renner, P.E.
Executive Director
Awwa Research Foundation

ACKNOWLEDGMENTS

The authors of this report wish to certain individuals and organizations for their contribution to this project:

The Pinellas County Utilities provided the project with water samples. Dr. Saleh Naser, University of Central Florida provided MAC strains, as did the Johne's Testing Center (JTC), University of Wisconsin. Varied test strains of bacteria were provided by the Wisconsin State Laboratory of Hygiene and Marty Collins of the same institution provided technical assistance.

Dr. Elizabeth Manning, University of Wisconsin, made substantive contributions to all facets of the project. The authors also thank John Albert, the AwwaRF project manager, and members of the project advisory committee for their support and thoughtful suggestions. Members of the project advisory committee were Matthew J. Arduino M.S., Dr.P.H., Audrey McDaniels, Ph.D and Giovanni Widmer, Ph.D.

EXECUTIVE SUMMARY

INTRODUCTION AND OBJECTIVES

Mycobacteria belonging to the *M. avium* complex (MAC) are opportunistic pathogens found in planktonic forms, in biofilms, and within protozoa. Members of this complex (i.e. subspecies *M. avium* ss. *avium* (MAA), *M. avium* ss. *paratuberculosis* (MAP), *M. avium* ss. *silvaticum* (MAS), *M. avium* ss. *hominisuis* (MAH) and, for the purposes of this endeavor, *M. intracellulare)* are environmental opportunistic mycobacteria commonly associated with human disease.

While a respectable amount of information is available in the literature about the occurrence of these pathogens, current methods of detection do not provide a sufficiently robust basis for conclusions about their true frequency and distribution in water distribution systems. Due to their environmental ubiquity, their suspected or known pathogenicity for humans and their ability to withstand standard water treatment protocols, MAC species are prime targets for improved methods of detection in finished water.

Specific objectives for this project included:

1. Evaluation of liquid culture methods for growth of MAC isolates exposed to stress conditions similar to those encountered in the environment and during water treatment processes.
2. Optimization of water filtration and elution methods.
3. Assessment of the effect of decontamination methods on the growth of MAC isolates and potential microbial contaminants in finished water.
4. Identification of anti-MAC antibodies most likely to be successful for immunomagnetic separation (IMS) concentration.
5. Evaluation of newly-developed IMS bead constructs for selective MAC concentration.
6. Establishment and evaluation of real time capacity for a validated multiplex MAC PCR.
7. Evaluation methods of detecting surface-associated and planktonic water habituated MAC.

METHODS

The ultimate goal of achieving a highly accurate and reliable MAC detection technique entailed optimizing numerous method components of a multi-step process. These steps included concentration of water samples; decontamination of the concentrated samples; selective concentration of mycobacteria; detection and enumeration of total MAC and identification of MAC species. Concentration approaches included centrifugation and filtration with a variety of filter types and centrifugation protocols. Decontamination trials assessed the efficacy of chlorine and cetylpyridinum chloride (CPC) against common water bacterial contaminants as well as mycobacteria. Since the number of mycobacterial cells in finished water samples is believed to be low, selective concentration is needed to ensure accurate mycobacterial species identification. Immunomagnetic separation techniques with both antibody-coated beads and antibody-coated processing vessels were evaluated. Methods for accurate identification and quantification of MAC species using liquid culture and genetic analysis (PCR) were validated. All steps are interrelated and were evaluated in conjunction with the steps that come before and after for effective optimization of a functional MAC detection and quantification procedure. The behavior of MAC species in annular reactors (as prototype water treatment equipment) using improved but still imperfect methods was investigated. New barriers to detection were identified and methods to remove these barriers were developed and assessed.

CONCLUSIONS

1. The method providing the greatest sensitivity in screening water for residual MAC species after treatment couples liquid culture-based methods with PCR. Combining the two approaches compounds their advantages while diluting their disadvantages. The culture-based method using the MGIT 960 system amplifies the number of organisms for subsequent DNA extraction, thus maximizing sensitivity while simultaneously limiting the effect of PCR inhibitors.

2. The MGIT culture method also provides a way to enumerate the MAC cells in a sample. This project developed the first algorithm to convert MGIT 960 instrument Time-To-Detection (TTD) values to Colony Forming Units (CFU) for any strain of MAP and has also established MAA strain-specific algorithms to convert TTD to CFU. The method has proven valuable as a research tool for rapid and cost-effective MAC enumeration. The utility of the method on finished water remains to be determined since the diversity of mycobacteria in some sample types could complicate MAC quantification by MGIT methods.

3. Use of free chlorine at standard concentrations to minimize contaminating non-mycobacterial organisms is unlikely to interfere with isolation of MAC species. Inclusion of the commonly used decontaminant CPC in water testing protocols may augment hydrophobic partitioning of mycobacteria and interfere with accurate MAC quantification.

4. Decontamination of water samples using CPC has little or no effect on MAC viability but substantially reduces the number of viable non-mycobacteria in samples enhancing their recovery by culture/PCR methods.

5. Filtration, followed by DNA extraction from the filter, is superior to centrifugation as a method of concentration and identification of MAC species from finished water samples but significant losses of MAC from samples occur due to hydrophobic interactions with filtration equipment, pipettes and/or other plasticware somewhat compromising the analytical sensitivity of assays.

6. Commercially available antibodies to MAC were not effective for immunoconcentration techniques but novel antibodies produced for the project were highly effective at selective concentration of MAA from water samples. However, technical problems prevented use of these antibodies with immunomagnetic beads. Alternative formats with greater surface area such as coated incubation trays are promising.

7. The triplex PCR method, which can be performed even in the presence of contaminating organisms, quickly identifies the presence of mycobacterial species in general as well as

specific MAC subspecies based on novel genetic targets and also provides a degree of quantification using real-time PCR technology.

8. Direct quantification of surface-associated MAC by placing coupons directly into culture media was superior to quantification with cells harvested from coupons by repeated cotton swabbing.

9. The hydrophobic properties of MAC have a significant impact on quantification methods, i.e., substantial numbers attach to plastic surfaces and seem to "disappear" from assays (see conclusion #4 above). The organisms' location and behavior in water systems, e.g., rapid attachment to solid surfaces, are dependent on the hydrophobicity of those surfaces. This fundamental finding needs further study as it significantly affects future MAC study designs.

10. Experiment protocols using single sample mini-chambers with direct-culture coupons for either static or moving water conditions permit unperturbed sampling, limit chances for contamination and minimize organism loss through handling.

CHAPTER 1
INTRODUCTION

INTRODUCTION

Background

The *Mycobacterium avium* complex (MAC) consists of *M. avium* (with subspecies *M. avium* ss. *avium* (MAA), *M. avium* ss. *paratuberculosis* (MAP), *M. avium* ss. *silvaticum* (MAS), *M. avium* ss. *hominisuis* (MAH) (Turenne et al., 2008) and, for the purposes of this endeavor, *M. intracellulare* (MI). These saprophytic organisms are acid-fast, slow to very slow-growing, non-pigmented rod-shaped bacteria, with complex, lipid-rich cell walls (Rastogi et al., 2001). *Mycobacterium avium* complex organisms are environmental opportunistic mycobacteria commonly associated with human disease {Abubakar, 2008; Feller, 2007; Turenne, 2007; Biet, 2005; Primm, 2004; Inderlied, 1993; Dorshorst, 2006 3449; Falkinham III, 1996; Falkinham, 2002; Greenstein, 2003; Hermon-Taylor, 2000; Kirschner, 1992; Leclerc, 2002}.

Human infections due to MAC appear to be increasing in developed countries and include three principal syndromes: cervical lymphadenitis in children, pulmonary infections in adults, and disseminated infection in AIDS patients (Bodle et al., 2008;Turenne et al., 2007). *M. avium* infects between 25 and 50% of AIDS patients in the United States and is a health threat to numerous other immunosuppressed populations (e.g., cancer patients). *Mycobacterium avium* ss. *paratuberculosis* has also been implicated in human disease and is suspected of causing a human gastrointestinal ailment (Crohn's disease) {Rubery, 2001; Hermon-Taylor, 2000; Grant, 2005; Feller, 2007; Chamberlin, 2001; Abubakar, 2008}. Infection by these organisms can be fatal. These pathogens are refractory to most antimicrobial agents due to their protective waxy wall and ability to persist within the host's immune cells (Hett and Rubin, 2008).

Mycobacteria have been isolated from home distribution systems, hot and cold water taps, ice machines, heated nebulizers, and showerhead sprays (Aronson et al., 1999;Falkinham et al., 2001;Hilborn et al., 2006;Le Dantec et al., 2002;Lehtola et al., 2007;von Reyn et al., 1994;von Reyn et al., 2002). Glover *et al.* (Glover et al., 1994) found that 34% of the mycobacterial isolates from the Los Angeles water supply were MAC organisms. MAC has also been found in natural water and in drinking water distribution systems throughout the United States at concentrations ranging from 0.8 to 45,000 per 100 mL (Falkinham et al., 2001). While water treatment reduced the number of mycobacteria in raw waters by 2 to 4 log units, high levels have been recovered from the water distribution system. Mycobacterial numbers were found to be 25,000-fold higher in distribution system samples than in those collected immediately downstream from the treatment facilities, indicating re-growth of those organisms surviving water treatment (Falkinham et al., 2001).

MAC organisms are also recovered in large numbers from water with low pH (optimal pH of 5.0 to 5.5), low oxygen concentration, zinc concentrations of around 0.75 g/L, and high organic matter content (Falkinham III, 1996; Kirschner et al., 1999; Kirschner et al., 1992). The positive association between *M. avium* CFU and raw source water turbidity suggests that *M. avium* cells are bound to colloidal or suspended particles (Falkinham et al., 2001). If adherent to particles of suspended solids, then the MAC content of raw water abstracted for domestic supply will be depleted by subsequent treatments, such as counter-current dissolved air floatation filtration. However, CT values (contact time concentration of chlorine i.e. disinfectant concentration ppm x time in minutes to reach 3 log units of cell death) for the effect of chlorine on MAP have been estimated to be up to 580 to 2,300 times greater than those for *E. coli* (Taylor et al., 2000; Whan et al., 2001). Gerba *et al.* (Gerba et al., 2003) concluded that "Current disinfection practices in the United States appear to be capable of dealing with most of the microorganisms on the Contaminant Candidate List, with the exception of *Mycobacterium avium*." If MAC elude the suspended solids treatment stage in water plants, they are unlikely to be destroyed by subsequent chlorination or ozone treatment (Taylor et al., 2000). In fact, Taylor found that water-grown *M. avium* cells were 10-fold more resistant than bacteriologic culture medium-grown *M. avium* cells and that resistance to disinfection is directly related to growth rate: the slower-growing strains are the most resistant (Taylor et al., 2000). They thus can accumulate in biofilms present in building (homes, hospitals, etc.) distribution systems.

Environmental MAC species may be taken up into protozoa (Falkinham III, 1996;Hermon-Taylor et al., 2000). *Mycobacterium avium* grown within *Acanthamoeba castellanii* appear to demonstrate an increased capacity to infect other amoebae, as well as macrophages and human colonic epithelial cells (Cirillo et al., 1997). Steinert et al. report that *M. avium* can survive for long periods within encysted forms of *Acanthamoeba polyphaga* (Steinert et al., 1998). Mycobacterial species, including MAP, have been reported as capable of entering a dormant phase during which time they are particularly resistant to antibiotics (Whittington et al., 2004).

Mycobacterium avium ss. paratuberculosis is the slowest growing strain in the MAC complex. In keeping with Taylor's disinfection findings (Taylor et al., 2000) our studies of MAC infections in animals have shown that MAP is the MAC species most resistant to antibiotics and disinfectants (Lambrecht et al., 1995;Rastogi et al., 1988;Shin and Collins, 2008). This organism causes a prevalent and fatal animal disease (Johne's disease - present in more than 70% of US dairy herds). Shed in manure at levels in excess of 10^8/gram of manure (and a cow produces over 30 kilograms of manure a day), infected animals heavily contaminate the environment and surface waters. MAP carries mycobacterial hardiness to a notable level: it has been recovered from the environment, still infective, more than a year after leaving the host (Shin and Collins, 2008;Whittington et al., 2003;Whittington et al., 2004). In addition to an association with Crohn's disease, its infection of AIDS patients has also been documented (Richter et al., 2002).

In summary, MAC species represent opportunistic pathogens that may persist in planktonic form, in biofilms, and within protozoa. While a respectable amount of information is available in the literature about the occurrence of these pathogens, current methods of detection do not provide a sufficiently robust basis for conclusions about their true frequency and distribution in water distribution systems. Due to their environmental ubiquity, their suspected or known pathogenicity for humans and their ability to withstand standard water treatment protocols, MAC species are prime targets for improved methods of detection.

OBJECTIVES

The overall aim of this study was to improve the methods used to detect MAC species in finished water, thus increasing utilities' confidence in their knowledge of the true prevalence of these organisms in finished water. The specific objectives of this study were to:

a. Evaluate liquid culture methods for isolation and quantification of MAC species.

b. Compare water filtration vs. centrifugation methods to isolate and concentrate MAC organisms from finished water samples.
c. Assess the effect of decontamination methods on the growth of MAC isolates and potential microbial contaminants in finished water.
d. Identify anti-MAC antibodies most likely to be successful for IMS concentration.
e. Evaluate newly-developed IMS bead constructs for selective MAC concentration.
f. Establish and evaluate real time capacity for a multiplex MAC PCR.
g. Evaluate methods under plant simulation for both surface-associated and planktonic MAC.

CHAPTER 2
METHODS COMMON TO MULTIPLE OBJECTIVES

The ultimate goal of achieving a highly accurate and reliable detection method will entail optimization of numerous method components, including: 1) sample collection (both biofilm and planktonic phases); 2) concentration of water samples and biofilms; 3) decontamination of the concentrated samples and/or; 4) selective concentration of mycobacteria (immuno-magnetic separation); 5) detection and enumeration of total MAC and identification of MAC species. The project required careful optimization of each individual component, an assessment of component permutations, and then validation of the best component combination.

STUDY STRAINS AND WATER SOURCES

Three MAC strains were used as standards throughout experimental trials: *Mycobacterium avium* ss. *avium* ATCC 35712, *M. avium* ss. *paratuberculosis* JTC303 (bovine clinical isolate from the Johne's Testing Center repository), and *M. intracellulare* ATCC 13209. Active cultures of all three strains were prepared frozen stocks and DNA was extracted for future use. *Mycobacterium avium* strains were cultured in 7H9 broth supplemented with 10% OADC (Becton Dickinson, Sparks, MD) for 2 weeks to 1 month at 37°C. For *M. avium* ss. *paratuberculosis,* 2 µg/ml of Mycobactin J (Allied Monitor, Fayette, MO) was added. Other mycobacterial and non-mycobacterial species used were obtained from the University of Florida and the Wisconsin State Laboratory of Hygiene.

Experiments requiring water-acclimated MAC entailed spiking strains described above in serial dilutions into autoclaved water from the University of Wisconsin pilot-scale drinking water treatment plant. Final testing focused on treated water samples provided by Pinellas County Utilities

ORGANISM QUANTIFICATION

Many experiments required a method to calculate the number of cells present in samples. Because mycobacteria are slow-growing and tend to clump, traditional plate-count methods can be insensitive. Experiments utilized two additional methods for quantification: flow cytometry and MGIT culture.

Flow Cytometry

The use of flow cytometry for enumeration of stock concentrations of microbial pathogens has been published (Hoffman et al., 1999). Briefly, an aliquot of the stock suspension is spiked with an FDA-approved, precisely enumerated microbead standard (FlowCount, Beckman Fullerton, CA) and analyzed by flow cytometry until 5000 microbeads have been counted. The number of organisms enumerated by light scatter is identified from the instrument's histogram and a ratio of organisms to microbeads is constructed. This ratio is multiplied by the lot-specific microbead concentration to determine the concentration of organisms in the stock suspension. In our initial trial, 50 ml cultures of *M. avium* ss. *paratuberculosis* JTC303 were centrifuged (2,000 x g, 30 min), washed and resuspended in sterile PBS + 0.05% Tween 80. The cell density of samples was quantified by flow cytometry as described above, giving $8.84 \pm 0.87 \times 10^7$ cells / ml (mean of triplicate samples ± standard deviation of the mean).

Plate-counting

This traditional method was employed to provide a comparison with liquid culture methods. Samples were removed just prior to liquid culture incubation, and either streaked directly or serially diluted and streaked to Middlebrook agar plates for CFU/ml determination. Agar plates were incubated at 37°C in 5% CO_2 and counts were made as soon as visible colonies appeared. Colonies were enumerated by visual counting.

MGIT 960 Liquid Culture

Serial dilutions of each mycobacterial strain were prepared and inoculated into MGIT media (Becton Dickinson, Sparks, MD) (Shin et al., 2007). Each tube contained 7 mL of medium and a fluorescent indicator embedded in silicone on the bottom of the tube. To each tube was added 800 µl of MGIT ParaTB supplement (Becton Dickinson, Sparks, MD), 500 µl of egg yolk suspension (Becton Dickinson, Sparks, MD), and 100 µL of VAN antibiotic cocktail (Becton Dickinson, Sparks, MD) resulting in final concentrations of 10 µg/mL vancomycin, 40 µg/mL amphotericin B, and 60 µg/mL nalidixic acid. Tubes were incubated at 37°C in a MGIT 960 instrument and removed when the instrument signaled them as being positive. The completed incubation tube has a headspace containing air which was partially displaced by the sample (typically 0.2 mL). The MGIT incubation machine tracked the inoculated tubes using Epicenter (a data management software program) that exported digital data hourly. MGIT 960 and Epicenter tracked the time to detection (TTD) for each MGIT tube. TTD is defined as the cumulative time from placement of the MGIT tube into the MGIT 960 instrument until a growth unit (GU) of 75 is reached. A growth unit represents the rate at which oxygen is being utilized within a MGIT tube. GUs accumulate until oxygen depletion, substrate depletion or inhibition occurs. (No negative controls accumulated a GU greater than 75 within 49 days during the study). The number of CFUs was then calculated based on TTD per a log linear standard curve shown in Appendix A.

Samples from all signal-positive tubes were subcultured on trypticase soy agar plates with 5% sheep blood (Becton Dickinson, Sparks, MD) to check for contamination. Acid-fast staining (Ziehl-Neelsen) was also performed on smears made from each signal-positive tube to confirm the presence of mycobacteria prior to identification by PCR.

LIVE/DEAD STAINING WITH FLOW CYTOMETRY

Several dyes, staining protocols, and detection platforms were evaluated for the ability to differentiate live from dead mycobacteria. Efforts primarily focused on the use of the BacLight® staining kit (Molecular Probes, Eugene OR). The BacLight® staining kit contains both green (Syto 9, excitation max 494 nm, emission max 515 nm) and red (propidium iodide, excitation max 536, emission max 617 nm) fluorescent nucleic acid dyes. When stained with a combination of these two dyes, all cells, regardless of viability status, will stain green while only cells in which the cell wall has been compromised will stain red. BacLight® has been used for enumeration of total and viable bacteria in drinking water (Boulos et al., 1999) and bacterial viability determination using this kit coupled with laser-based detection systems (flow cytometers) has been reported (Gasol et al., 1999;Langsrud and Sundheim, 1996). The use of this kit for mycobacteria staining has been described (Barker et al., 1997).

CHAPTER 3
OBJECTIVE ONE

OBJECTIVE 1: EVALUATE LIQUID CULTURE METHODS FOR ISOLATION AND QUANTIFICATION OF DIFFERENT SPECIES OF *MYCOBACTERIUM AVIUM* COMPLEX.

Quantification of viable bacteria is the foundation of innumerable studies such as MAC environmental survival, resistance to pasteurization, or resistance to decontamination agents. This research requires precise and sensitive quantification methods for viable cells. For the slowly growing MAC species that have a strong tendency to clump, this seemingly simple task can be difficult, expensive and imprecise. Organism clumping interferes with accurate counting of colony forming units (CFUs) if traditional solid media plate protocols are used. Standard plate counting methods for slowly growing mycobacteria also fail due to either contamination or solid medium dehydration during the 30-50 day incubation period. Spectrophotometric methods for standardization of cell suspensions provide only rough estimations of bacterial cell concentrations and show more than 1 $\log_{10}$ CFU/ml variations (Harris and Barletta, 2001;Reddacliff et al., 2003;Shin et al., 2007).

Liquid (vs. solid) media methods produce culture results more swiftly and offer the possibility of equal to greater quantification accuracy. A new generation of liquid culture system (BACTEC MGIT 960, Becton Dickinson, Sparks, MD) has replaced the BACTEC 460's radioactive detector and included automated culture incubation, reading, and reporting of positive samples. This system was originally designed for detection of mycobacterial species commonly found in human clinical samples. It uses an oxygen-quenching fluorescent sensor technology in conjunction with software algorithms to determine when tubes are "positive", i.e., when significant mycobacterial growth has occurred. The newly developed media (MGIT ParaTB medium Becton Dickinson, Sparks, MD), supports growth of multiple MAC species and subspecies, including the most fastidious and slow-growing ones, and the incubator interprets fluorescence measurements made hourly on each culture tube (Cruciani et al., 2004;Garrigo et al., 2007;Hillemann et al., 2006;Huang et al., 2001;Kruuner et al., 2006;Somoskovi et al., 2000). The instrument reports results as "Time To Detection" (TTD) in days and hours. Supplemental testing, usually by acid-fast staining and PCR, is required to confirm the identity of mycobacterial isolates (See Objective 6).

The goal of this objective was to determine if TTD correlated with MAC species' CFU to provide an accurate method for counting MAC cells.

Specific Aims For This Objective Were:

1. Derive a relationship between MAC inoculum cell number and the Time To Detection (TTD) signal in MGIT liquid culture tubes
2. Use the BACTEC daily growth index values to determine if growth equation coefficients are the same for different MAC species.

METHODS

Experimental design – Phase I. Counting experiments were completed with MAA ATCC 35712, MAA EPA-1 and MAP ATCC 19698. In these experiments, log-phase cells were washed, resuspended in PBS buffer, and homogenized by vortexing to dissipate cell clumps and then sequentially filtered through 8.0, 5.0 and 3.0 µm filters to remove residual clumps (Sung and Collins, 1998). The resulting single-cell suspensions were serially diluted and the number of cells in each dilution determined by multiple methods with plate counting on solid media being the designated reference method.

Experimental design – Phase II. A more detailed analysis of variance among the MAP subtype of MAC isolates was conducted focusing on the assay with the broadest dynamic range and analytical sensitivity, i.e. MGIT method. Multiple MAP strains were evaluated using single-cell suspensions of MAP to compare plate counts (the CFU reference method) to MGIT TTD results.

Live/Dead staining with flow cytometry. The BacLight® kit was evaluated as a means of rapidly determining the impact of various water sample decontamination procedures on mycobacterial viability and to quickly confirm initial seed culture inoculum densities.

ATP assay. The BacTiter-Glo® ATP assay (Promega, Madison, WI) is a new quantitative ATP assay for bacteria. The assay contains a proprietary cell lysing buffer and chemically stable luciferase and luciferin reagents. This reagent is added to cell samples in a microplate and relative light units (RLU) are measured on a luminometer. In the assay, ATP standards are included to allow conversion of RLU to amount of ATP. This assay had not been validated for mycobacteria and given the organisms' thick, lipid-rich cell wall it was necessary to assess the cell wall lysis capability of BacTiter-Glo® buffer. Unless otherwise noted, the kit was used according to manufacturer's instructions (Promega, 2008).

BACTEC 460. The growth index (GI) of each vial of BACTEC modified 12B medium inoculated with single-cell suspensions of MAC were monitored daily for 30 days. The (CGI) was then used to determine growth equation coefficients needed for enumeration based on a previously published MAC growth model (Lambrecht et al., 1988).

MGIT 960. For Phase I studies the TTD of each tube of supplemented MGIT ParaTB Medium inoculated with each serial dilution of MAC was recorded over a 49 day incubation protocol. The relationship of MGIT TTD versus CFU for each MAC strain determined by plate counting was analyzed by linear regression (GraphPad Prism version 5.00 for Windows, GraphPad Software, San Diego California USA, www.graphpad.com). For Phase II studies a larger number of MAC strains of the MAP type were tested and multiple regression methods were evaluated to determine the regression model best fitting the data.

RESULTS

Flow cytometry. MAC cell counts by flow cytometry were linear with CFU determined by plate counting for both MAA and MAP; $R^2 = 0.98$ (Figure 3.1). The analytical sensitivity of flow cytometry was roughly 10^4 MAC/mL. For MAA these results were consistent in duplicate independent experiments (Table 3.1).

Table 3.1 *M. Avium* ATCC 35712 Plate Counts Vs. Flow Cytometry (Two Trials)

	Trial 1		Trial 2	
Dilution	Plate counts[1]	Flow cytometry	Plate counts[2]	Flow cytometry
10^{-1}	1.4×10^6	$6.4 \pm 0.6 \times 10^5$	3.7×10^6	$7.7 \pm 0.7 \times 10^5$
10^{-2}	1.4×10^5	$7.6 \pm 0.6 \times 10^4$	3.7×10^6	$9.7 \pm 0.7 \times 10^4$
10^{-3}	1.4×10^4	$9.7 \pm 0.6 \times 10^3$	3.7×10^6	$7.7 \pm 0.7 \times 10^4$
10^{-4}	1.4×10^3	below detection	3.7×10^6	$7.7 \pm 0.7 \times 10^3$
10^{-5}	1.4×10^2	below detection	3.7×10^6	below detection
10^{-6}	1.4×10^1	below detection	3.7×10^6	below detection
10^{-7}	1.4×10^0	below detection	3.7×10^6	below detection

[1]Based on CFU for 10-5 dilution plates = 140 ± 44 CFU/plate
[2]Based on CFU for 10-5 dilution plates = 37.8 ± 7.2 CFU/plate

Live-dead staining with flow cytometry results. Flow cytometry was used to enumerate live and heat-inactivated MAA. In these experiments, live and heat-killed (2 hr at 80°C) MAA were mixed such that suspensions contained 5×10^5 total MAA cells total at various live:dead ratios. Organisms were stained for 15 minutes at room temperature with a 1:1000 dilution of the stock Syto 9 solution provided in the BacLight® kit and propidium iodide prepared from a stock solution (5 μg/mL final concentration). The estimated live:dead ratio of MAA cells using the BacLight® kit closely approximated the predicted live:dead MAA cell ratio. (It should be noted suspensions of untreated actively growing cultures of MAA contain a small percentage of dead cells and thus results will differ slightly from the target concentration). (Table 3.2). The vital dye staining method however was not used to determine killing in disinfection studies since the lower limit of detection was significantly higher than the MGIT culture system.

Table 3.2. Flow Cytometry Results for Baclight Live/Dead Staining of Mixtures of Live and Heat-Killed MAA ATCC 35712, MAP DT 303, MAP ATCC 19698 (Triplicate Trials)

Target Live:Dead Ratio	***MAA* ATCC 35712 Observed Live:Dead Ratio (SD)**	***MAP* DT 303 Observed Live:Dead Ratio (SD)**	***MAP* ATCC 19698 Observed Live:Dead Ratio (SD)**
0:100	1:99 (0.12)	0:100 (0.60)	7:93 (0.55)
10:90	12:88 (0.92)	10:90 (0.17)	18:82 (1.14)
25:75	23:77 (0.86)	32:68 (0.84)	30:70 (0.65)
50:50	46:54 (0.71)	46:54 (0.58)	43:57 (0.65)
75:25	69:31 (1.48)	68:32 (0.47)	57:43 (0.76)
90:10	83:17 (0.68)	81:19 (0.75)	66:34 (0.68)
100:0	94:60 (1.76)	92:8 (0.36)	77:23 (2.21)

ATP assay results. ATP measurements from MAC cells were possible using the BacTiter-Glo® assay (picomoles ATP). The lower limit of detection was in the range of 10^3 to 10^4 MAC cells per 100 μL, significantly higher than limit of detection reported for other bacterial species i.e. <100 cells / sample (Promega, 2008). Hypothesizing that lysis of MAC was not complete because of its thicker cell wall thus compromising the analytical sensitivity of the ATP assay, several alternative methods for MAC cell lysis were explored.

The addition of a Tris-EDTA buffer followed by boiling for 1 min significantly decreased the RFU signal compared to untreated cells resuspended in PBS for both MAA and MAP (data not shown). For cells resuspended in PBS, boiling for 1 min also decreased the RFU signal compared to untreated cells. Freezing the cells at –80˚C and thawing just prior to the assay increased the RFU signal compared to cells not frozen (data not shown). In subsequent trials with the ATP assay the freeze-thaw procedure was added as an additional lysis step.

Using the serially diluted samples from the counting experiments described above, the amount of ATP measured was linear ($r^2 < 0.99$) with respect to cell number between 10^6 and 10^3 cells for both MAA and MAP (Figure 3.2). The limit of detection was roughly 10^3 cells / 100 µl sample. In a separate experiment, we tested the effectiveness of the BacTiter-Glo® assay buffer to lyse (kill) MAC cells. Cell suspensions of MAA and MAP were frozen at –80˚C for 30 min, thawed at room temperature, and mixed with an equal volume of BacTiter-Glo® assay buffer. After a five minute incubation with the buffer, the mixture was inoculated to MGIT culture tubes to assess the number of live cells still present. There was a significant decrease ($p < 0.01$) in the number of viable cells after treatment with the BacTiter-Glo® assay buffer for both MAA and MAP (Table 3.3). However, the percent decrease was only 50% and 64% for MAA and MAP, respectively, indicating the majority of cells remained viable after treatment. This result may explain the relatively poor analytical sensitivity of the assay for MAC species as compared to other bacteria.

Table 3.3. Effect of Freezing at –80˚C and Promega Bactiter-Glo® Lysis Buffer on the Viability of MAA and MAP. Data are Mean ± Standard Deviation or Three Replicate Samples.

	Treatment	Time to MGIT Detection (TTD; days)	TTD converted to cell number*
MAP	Untreated	8.17 ± 0.07	$2.26 \pm 0.10 \times 10^5$
	Freeze	8.01 ± 0.23	$2.50 \pm 0.36 \times 10^5$
	Freeze+buffer	9.81 ± 0.27	$8.19 \pm 1.42 \times 10^4$
MAA	Untreated	5.14 ± 0.02	$6.48 \pm 0.19 \times 10^4$
	Freeze	5.40 ± 0.02	$4.66 \pm 0.14 \times 10^4$
	Freeze+buffer	5.69 ± 0.09	$3.25 \pm 0.34 \times 10^4$

*MAP: log cell number = -0.2714(TTD) + 7.5706. MAA: log cell number = -0.5425(TTD) + 7.5995

BACTEC 460. This liquid culture system, with its enumeration algorithm established for MAP, was adapted for use with MAA species in these experiments using growth data to derive algorithm regression coefficients specific to these mycobacterial species. The growth data from samples containing 10^2 to 10^6 cells per bottle provided the best fit ($r^2 = 0.905$) in the regression analysis.

MGIT 960. For both MAA and MAP, the TTD signal recorded by the MGIT 960 instrument was linear ($r^2>0.983$) with respect to the log cell number as measured by plate counts from 10^1 to 10^6 cells per tube and reproducible when comparing independent experiments for both MAA and MAP (Figure 3.3). The lower limit of detection in this system was fewer than ten cells per tube.

Although the relationship between TTD measurements by MGIT and CFU measured by plate counting was linear and reproducible there was significant variance among MAA strains, more so than for MAP strains. Representative results are shown in Figure 3.4.

Phase II studies thus sought to determine a single MGIT counting algorithm for the more pathogenic of MAC subtypes, namely MAP, and to elucidate the sources of algorithm variation. A total of 12 MAP strains were evaluated: ATCC19698, K-10, JTC100, JTC102, JTC114, JTC303, USF-5, UCF-7, B213, B236, B238, and B244. Strains ATCC19698 and K-10 strain are type strains and those with the JTC prefix are clinical isolates recovered from bovine fecal or tissue samples by the Johne's Testing Center (School of Veterinary Medicine, University of Wisconsin-Madison). Strains UCF-5 and UCF-7 are human-origin MAP recently isolated from Crohn's disease patients. These and 4 bovine strains, B213, B236, B238, and B244, were kindly provided by Dr. Saleh A. Naser (University of Central Florida, Orlando, FL). Serial dilutions of single cells suspensions of each *M. paratuberculosis* strain were prepared and 100 μL was inoculated into ParaTB MGIT medium. This experiment was repeated at least twice and MGIT ParaTB tubes were inoculated in triplicate. The TTD (days) of each tube was recorded and plotted versus $\log_{10}$ CFU inoculum to generate a standard curve for each strain tested; referred to as a strain-specific standard curve. In addition, the data for all 12 *M. paratuberculosis* strains was pooled to create generalized standard curve. The model best fitting the data was determined using statistical software (GraphPad Prism version 5.00 for Windows, GraphPad Software, San Diego California USA, www.graphpad.com).

The validity of the generalized standard curve for each of 12 strains was determined by cross-validation and prediction errors. In the cross-validation CFU counts for each of the 12 strains were within the 95% prediction intervals regardless of cell concentration when compared with the generalized standard curve based on the other 11 strains. As expected, strain-specific curves were more accurate for that strain's enumeration than was the generalized standard curve with very narrow 95% prediction intervals.

The effect of inoculum volume, with number of MAP cells held constant, was tested over a wide range of dilutions. Ten-fold serial dilutions of 3 MAP strains were prepared as described. MGIT ParaTB tubes were inoculated with either 100 μL or 1,000 μL of MAP cell suspension. Predicted CFU counts for tubes with different inoculum volumes but the same expected number of MAP cells, e.g., 100 μL of 10^5 CFU/mL versus 1,000 μL of 10^4 CFU/mL, were compared. Counts were highly correlated ($r>0.99$) regardless of inoculum volume.

The TTD reported by the MGIT 960 system was highly correlated to colony counts determined by standard plate counting methods. Standard curves for all 12 strains approximated a linear relationship between CFU and TTD but the data best fit a one-phase exponential decay curve mode. While the greatest accuracy in predicting MAP CFUs was obtained by use of a strain-specific standard curve (0.31 $\log_{10}$ CFU, 95% PI), it was possible to pool data from 12 MAP strains to create a generalized standard curve with narrow prediction errors (<1 $\log_{10}$ CFU). (See Appendix A for standard curve formula).

Liquid culture : MGIT vs. BACTEC. Using only the ATCC19698 strain of MAP, CFU predictions were compared using the BACTEC model and MGIT generalized standard curve (triplicate determinations). While the predicted CFU counts were the same for both systems for MAP at $10^{3.49}$ to $10^{6.49}$ (Multiple comparison test, $P=0.88$, $r>0.99$), results were obtained in less than half the time and over a greater dynamic range (6 logs vs. 3 logs) for the MGIT method as compared to the BACTEC method.

DISCUSSION

The MGIT method was superior to other MAC counting methods (Table 3.4). Counting MAC by flow cytometry was rapid, simple and correlated well with plate count CFUs but the analytical sensitivity of the flow cytometry method was 10^4 CFU/mL limiting its utility. Counting MAC by using ATP as a biomarker was similarly good at quantifying moderate to high MAC cell numbers but suffered from poor analytical sensitivity, i.e. $10^3 - 10^4$ MAC cells. An algorithm for conversion of BACTEC 460 results as CGI to CFU showed poor agreement between plate count CFUs BACTEC MAC algorithm count estimates with MAA species even though prior studies with MAP showed good agreement of BACTEC CGI and plate count CFU (Lambrecht et al., 1988). Given these results, neither flow cytometry, ATP measurement or the BACTEC 460 system were included as MAC enumeration methods for subsequent elements of the project.

The MGIT 960 counting method (conversion of TTD to CFU by a prescribed standard curve) was the best counting method of those compared in this study with the best analytical sensitivity (~10 MAC cells) and a reproducible one-phase exponential decay curve relationship between TTD and CFU (r^2>0.96). Standard curves for conversion of TTD to CFU for MAA were strain-specific and used as such in subsequent experiments requiring MAA quantification. Standard curves for MAP were less variable and universal curves were established for laboratory-adapted strains and low-passage clinical strains (Shin et al., 2007). The ability to define universal standard curves for MAP strains (curves able to fit multiple MAP strains) was likely due in part to their much slower growth rate compared to MAA thereby lessening the variance due to instrument culture reading frame.

Table 3.4 Characteristics, Advantages, And Disadvantages of MAC Counting Methods

Method	Plate Count	BACTEC	MGIT	Flow Cytometry	ATP
Dynamic range	25 – 250 CFU/plate	$10^3 - 10^7$ cells/mL	$10^1 - 10^7$ cells/mL	$10^4 - 10^7$ cells/mL	$10^5 - 10^8$ cells/mL
Time to result	>30 days	15-30 days	3-30 days	Minutes	<1 hour
Labor	Moderate	High	Low	Low	Low
Advantages	"Gold standard"	Well established for MAP Good dynamic range	Automatic culture incubation and monitoring Wide dynamic range	Rapid and easy	Rapid and easy Provides cell viability measurement
Disadvantages	Affected by cell clumping Long time to get results Resource intensive	Requires daily manual culture reading Radioactive culture media (^{14}C)	Cost ~ \$10/sample	Affected by cell clumping Limited dynamic range Cannot distinguish live from dead cells	Limited dynamic range Poor cell lysis efficiency (requires more study for optimization)

The MGIT system provides a useful, accurate, and precise method for quantification of viable MAC cells that avoids the problems seen by traditional plate counts (clumping and long incubation) and the BACTEC 460 counting algorithm (reliance on ^{14}C-containing compounds that must be manually monitored daily). It is also relatively low cost in both labor and materials and highly reproducible between assays such that duplicate measurements were sufficient to provide excellent MAA counting precision. The limitation of the counting method is that it requires essentially pure cultures. Enumeration of MAC in finished water samples by MGIT methods will require sample preparation methods that remove or inactive contaminating microflora.

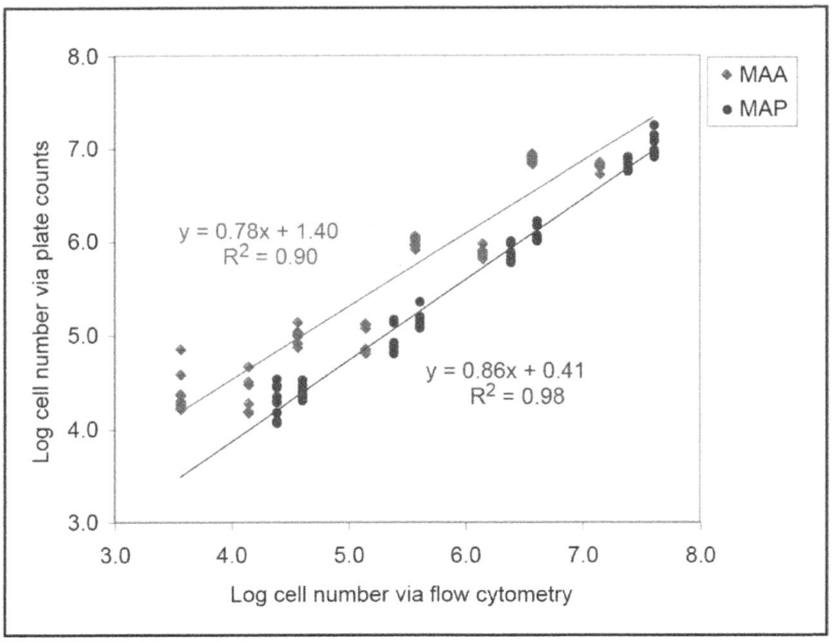

Figure 3.1. Flow Cytometry Counting Results Compared to Plated Count Cfus for MAA and MAP.

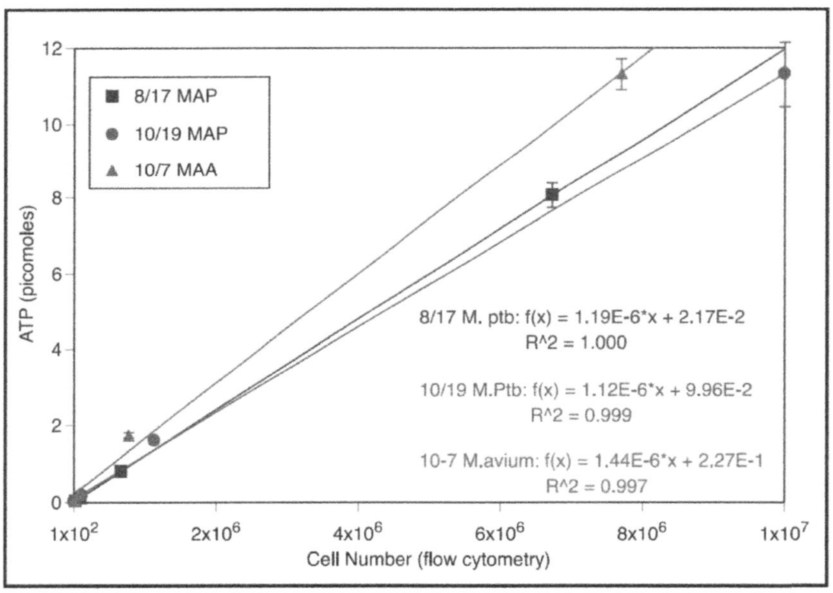

Figure 3.2. Comparison of ATP Results with Flow Cytometry Counts for MAA and MAP.

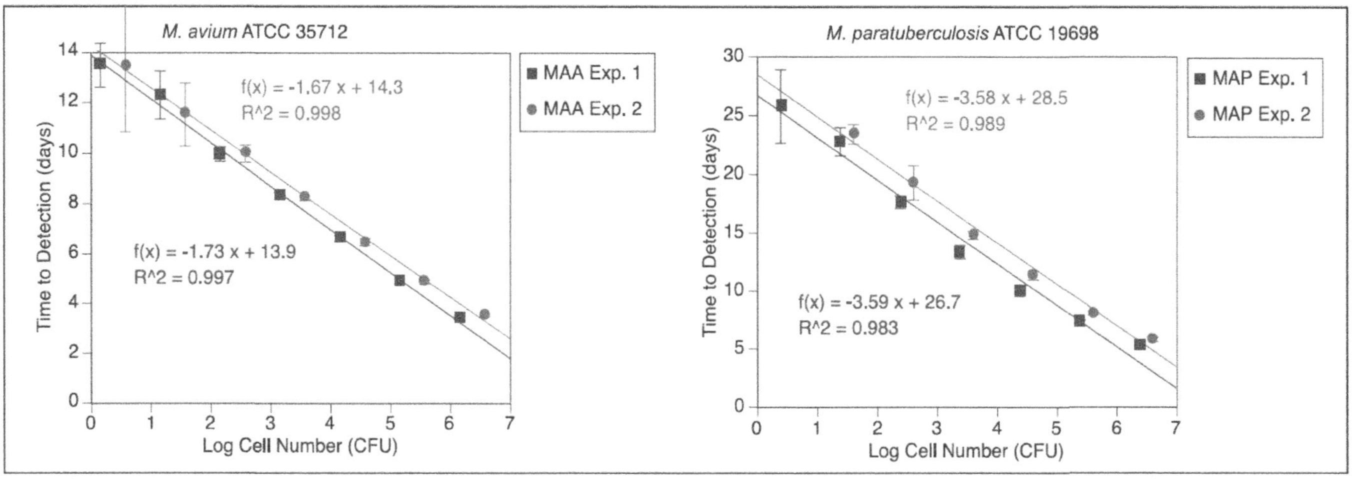

Figure 3.3. Reproducibility of MGIT TTD Measurements Over A Wide Range of Cells Counts (CFU) for MAA ATCC 35712 and MAP ATCC 19698 for Two Independent Trials.

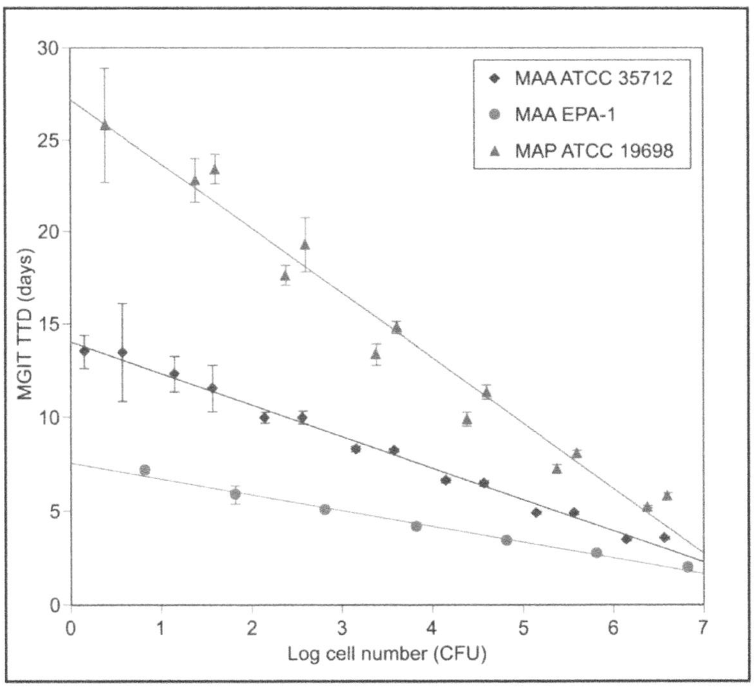

Figure 3.4. Relationship of MGIT TTD and Plate Counts (CFU) for MAA ATCC 35712, MAA EPA-1, and MAP ATCC 19698C.

CHAPTER 4
OBJECTIVE TWO

OBJECTIVE 2: COMPARE WATER FILTRATION VS. CENTRIFUGATION METHODS TO ISOLATE AND CONCENTRATE MAC ORGANISMS FROM FINISHED WATER SAMPLES

A series of experiments were conducted to optimize filtration and centrifugation methods for detection of MAC in water. While the experiments were done in succession with the results of each experiment guiding the design of subsequent experiments, the methods, results and discussion are combined for greater clarity. The details of the methods can be found in Appendix one.

METHODS AND MATERIALS

MAC strain MAA ATCC 35712 was used as a model organism to spike drinking water and then assess recovery efficiency. MAC recovery was assessed using MAC cell counts based on the MGIT counting method established in Objective #1. Additionally, MAC concentration methods were evaluated based on total recoverable DNA and MAC-specific DNA by semi-quantitative real-time-PCR methods concurrently under development (Objective #6).

Trials addressed the effectiveness of:

- Centrifugation
  - 1a -Assessment of centrifugation force and duration on recovery of MAC
- Filtration
  - 2a - Sonication
  - 2b - Acetone treatment
  - 2c - Pore size
  - 2d - Filter type, filter break-through
  - 2e - MAA attachment to glass
  - 2f - Sigmacote™ treatment of glass bottles

Trial 1a: test the effect of *centrifugation force and duration* on MAA recovery. The inoculated bottles were centrifuged at one of three speeds (2830 x g, 4420 x g, or 12800 x g) for either 30 or 60 min. After centrifugation, the supernatant was decanted and the pellet was resuspended by vortexing in 5 ml PBS buffer. Three, 100µl aliquots of each resuspended pellet were inoculated to MGIT tubes of ParaTB medium to determine the number of MAA cells recovered.

Trial 1b: test the effect of using *fewer sample containers for centrifugation.* High speed centrifugation (12,800 x g for 25 min) was employed using a method of water processing that minimized the number of centrifuge bottles. As an additional measure of MAA recovery total DNA was quantified using the Quant-iT™ High Sensitivity DNA Assay Kit (Invitrogen, Carlsbad, CA).

Filtration. These trials employed autoclaved Madison tap water inoculated with approximately 1×10^6 MAA cells, vigorously shaken and then filtered. The filters were placed directly into MGIT tubes to determine the number of MAA cells captured on each filter.

Trial 2a focused on *recovery of MAA whole cells from PCTE filters by sonication.* The filters were sonicated in a bath sonicator (Branson, 60 Hz, 117V) for 5 or 10 minutes. (Previous experiments showed that this manner of sonication did not affect the viability of MAA).

For Trial 2b, *acetone treatment of filters,* dissolvable cellulose acetate filters that have been used to recover *Cryptosporidium* oocysts from water (Aldom and Chagla, 1995) were employed. The sensitivity of MAA to acetone and ethanol was tested by exposing the organisms to the compounds and then determining the number of viable cells remaining.

To test the effect of PCTE *filter pore size* in MAA recovery for **Trial 2c,** each bottle of spiked water was filtered through either a 0.4 μm or 0.2 μm pore size filter. Since a plausible explanation for loss of MAA cells during filtration concentration experiments is passage of the MAA cells through the 0.4 μm filter, i.e. filter break-through, aliquots were taken from the filtrate as well to determine the number of organisms that may have passed through the filter.

For Trial 2d, four *different filter types* were compared: polycarbonate track etched filter (PCTE); mixed cellulose acetate cellulose nitrate filters (cellulose, Fisher Scientific 09-719-2E); nylon (Whatman 7404-004); and Durapore™, a hydrophobic polyvinylidene fluoride membrane (Millipore, HVLP 04700). All filters tested were 0.4 μm pore size.

To assess relative DNA recovery (vs. whole cell recovery described above) from different filter types, the filters themselves were transferred to ribolyser tubes for DNA extraction. In addition, inoculated samples of PBS in glass bottles were tested for viable MAA, total DNA and IS*1311* both before filtration (MAA cells in suspension) and after filtration (MAA cells trapped on the 0.4 μm PCTE filters). As in Trial 2c, the filtrate (water that passed through the filter) was sampled to determine if MAA cells were breaking through the filter.

Trial 2e investigated the *adherence of MAA cells to glass.* The base of the glass filter bell was swabbed with a sterile cotton swab and the swab transferred directly to a tube of MGIT ParaTB medium to assess the number of MAA cells adherent to the glass filter bell.

Trial 2f tested a *blocking agent* to minimize MAA cell loss since prior filtration trials had significant, i.e. 80% or more, loss of MAA cells. Sigmacote™ (Sigma-Aldrich, St. Louis, MO), a silicone solution in heptane that forms a covalent microscopically thin film on glass, was investigated as an approach to block MAA cells from adhering to surfaces in contact with the MAA-containing water.

Having finished these concentration trials, the optimized methods were applied to treated water samples kindly provided by Pinellas County Utilities. (Again, details for the methods used in this Chapter can be found in Appendix A.)

RESULTS

Significant losses of MAA whole cells spiked to tap water occurred during processing for both centrifugation and filtration. (Table 4.1). MAA concentration by centrifugation resulted in a loss of roughly 2 logs of MAA cells (99.8%). MAA concentration by filtration results in a loss of approximately 1 log (~90%) of MAA cells.

Table 4.1. Centrifugation Vs. Filtration for The Recovery of Whole Cell MAA.

Sample (spiked tap water)	Total cell number Pre-process	Total cell number Post-process	% loss
Centrifugation Rep 1	$6.61 \pm 2.40 \times 10^5$	$1.04 \pm 0.31 \times 10^3$	99.8
Centrifugation Rep 2	$1.59 \pm 0.74 \times 10^6$	$3.63 \pm 0.83 \times 10^3$	99.8
Centrifugation Rep 3	$2.11 \pm 1.45 \times 10^6$	$5.40 \pm 0.16 \times 10^3$	99.7
Filtration Rep 1	$2.06 \pm 1.01 \times 10^6$	2.55×10^5	87.6
Filtration Rep 2	$1.97 \pm 1.56 \times 10^6$	1.52×10^5	92.3
Filtration Rep 3	$1.25 \pm 0.41 \times 10^6$	1.37×10^5	89.1

In another trial coupling MAA whole cell with MAA DNA recovery (Table 4.2), comparable losses were noted for whole cell numbers. Recovery of DNA from centrifuged samples was below the limit of assay detection (<0.02 ng/µl). It should be noted, however, that due to the limited capacity of the ribolyser tubes used for DNA extraction, only 0.3 mL of the 5.0 mL total final volume was used for DNA extraction. DNA recovery from the filtered samples was significantly higher at 46.8 ng DNA (Table 4.2) since the entire sample trapped on the filter was used for DNA extraction. The DNA isolated from these samples was used in the real-time PCR assay (see Objective # 6) using standard PCR conditions (5 µl samples per reaction and 65°C annealing temperature). The DNA extracted from both the centrifuge- and filter-concentration methods gave positive PCR results using IS*1311* primers. The crossing threshold (Ct) was higher for the centrifuged samples indicating lower PCR target (IS*1311*) numbers, consistent with the lower total DNA concentration.

Table 4.2. Centrifugation Vs. Filtration for Recovery of MAA Whole Cells Vs. DNA.

Spiked tap water	Initial / 1L	Total cells recovered	% loss	DNA Isolated (ng)	PCR results (IS*1311*/mean Ct)
Centrifugation	$2.1 +/- 0.7 \times 10^7$	$1.1 +/- 0.3 \times 10^6$	94.6	Below detection	Positive / 42.5
Filtration	$1.4 +/- 0.4 \times 10^7$	$1.1 +/- 0.3 \times 10^6$	92.2	46.8 per 100ul	Positive / 28.5

Trial 1a sought to improve centrifugation results by increasing the centrifugation speed to 12,800 x g (near the maximum velocity for the available centrifuge) improved recovery of MAA from tap water from 1% at 2,830 x g for 25 min to 20.7% (average of replicate experiments) at 12,800 x g for 60 min (Table 4.3).

Table 4.3 Trial 1a-effect of centrifugation speed and Time for Recovery of whole Cell MAA

Centrifugation conditions	Total cell number before centrifugation[1]	Total cell number after centrifugation[2]	% loss
2,830xg, 25 min	$2.33 \pm 0.41 \times 10^5$	$2.25 \pm 0.73 \times 10^3$	99.0
4,420xg, 25 min	$2.33 \pm 0.41 \times 10^5$	$8.01 \pm 5.29 \times 10^3$	96.6
4,420xg, 60 min	$2.33 \pm 0.41 \times 10^5$	$1.62 \pm 0.75 \times 10^4$	93.1
12,8000xg 25 min	$2.33 \pm 0.41 \times 10^5$	$5.14 \pm 0.33 \times 10^4$	78.0
12,8000xg 60 min	$2.33 \pm 0.41 \times 10^5$	$4.52 \pm 1.18 \times 10^4$	80.6

[1]Total number of cells in 200 ml calculated from average of the initial 2 L of spiked tap water
[2]Total number of cells in 5 ml, average of 2 replicate samples and triplicate MGIT counts

An initial approach used to assess filtration focused on whole cell recovery (viable counts) and DNA extraction efficiency from these PCTE filters using PBS vs. tap water. (Table 4.4). The number of total MAA cells suspended in PBS was not significantly different from the number estimated trapped on the PCTE filter ($p>0.05$). Total DNA extracted from MAA cells suspended in PBS was not significantly different from the number estimated trapped on the PCTE filter ($p>0.05$). And, IS*1311* real-time PCR Ct values for suspended MAA cells was not different from that found trapped on PCTE filters ($p>0.05$).

Table 4.4 Viable MAA Cells, Total MAA DNA And IS*1311* Real-Time PCR Results for Filtered MAA Spiked in PBS Buffer.

Sample type	Cell number	DNA Isolated (ng)	PCR results (IS*1311* / mean Ct)
Cells in suspension	$2.7 +/- 0.1 \times 10^6$	43.4 +/- 1.1	Positive / 28.2
Cells on filter	$1.1 +/- 0.3 \times 10^6$	46.8	Positive / 28.5

Given these encouraging results, filtration trials focused on increasing MAA recovery through filter sonication. For Trial 2a, the majority of the cells remained adhered to the filter after sonication. There was no significant difference ($p<0.01$) between the number of cells eluted from the filter with 5 or 10 min sonication as shown in Table 4.5.

Table 4.5. Trial 2a - Elution of MAA Cells from PCTE Filters using Sonication.

Sonication Time	Rep	Initial Cell count	Filter cell count After sonication	Sonication buffer Cell count *
5 min	1	$3.00 \pm 1.67 \times 10^7$	1.75×10^6	$3.84 \pm 0.30 \times 10^5$
	2	$1.92 \pm 0.14 \times 10^7$	5.87×10^5	$9.45 \pm 1.09 \times 10^4$
10 min	1	$2.36 \pm 0.00 \times 10^7$	8.90×10^5	$3.82 \pm 3.35 \times 10^5$
	2	$2.76 \pm 0.20 \times 10^7$	8.45×10^5	$6.08 \pm 1.26 \times 10^4$

*Total number of cells in the 5 ml sonication buffer

Trial 2b (no data table) assessed the effect of acetone on MAA in anticipation of employing acetone-dissolving filters. The results were definitive: all three MGIT cultures inoculated with acetone-treated MAA cells were negative after 45 days of incubation at 37°C. In contrast, MAA cells treated with ethanol only gave MGIT TTD values similar to those of the untreated MAA controls. Acetone was thus shown to be lethal for MAA cells.

Filter pore size was evaluated as a factor for MAA recovery. There was no significant difference ($p<0.01$) in MAA cell recovery between the 0.4 and 0.2 μm PCTE filters. (Table 4.6.) Approximately 1 log of MAA cells was lost in the filtration procedure. The MGIT counts from the filtration effluent were not consistently positive after 25 days of incubation (data not shown) indicating the cell number in the effluent was less than 1 cell / 200μl. MAA cell loss was possibly due to passage through the filter or adherence to sample bottle or glass filter bell. The filtration rate of tap water through the 0.4 μm filter was noticeably faster then through the 0.2 μm filter.

Table 4.6. Trial 2c - Recovery of MAA with Different Pore Size Filters

Filter	Rep	Initial Cell count	Filter Cell count	Filter count Average
0.4 μm	1	$2.13 \pm 0.16 \times 10^7$	1.94×10^6	
	2	$2.63 \pm 0.39 \times 10^7$	1.94×10^6	
	3	$1.42 \pm 0.99 \times 10^7$	9.38×10^5	$1.61 \pm 0.58 \times 10^6$
0.2 μm	1	$2.36 \pm 0.16 \times 10^7$	8.90×10^5	
	2	$1.75 \pm 0.38 \times 10^7$	6.86×10^5	
	3	$6.88 \pm 7.02 \times 10^7$	7.23×10^5	$7.66 \pm 1.09 \times 10^5$

Trial 2d reviewed different types of filters for both cell and DNA recovery. There was no significant difference ($p>0.05$) in the number of viable MAA cells recovered with PCTE, mixed cellulose, or Durapore™ membranes (Table 4.7a). Insignificant numbers of cells were lost by breaking through the filter with this method (fewer than one MAA cell in 300 μl).

Table 4.7 A. Trial 2D - Recovery oF Maa Cells Using Different Pore Size Filters

Experiment	Filter type	Total cell counts initial (average)†	Total cell counts on filter*	% loss
A	PCTE	$2.1 \pm 0.8 \times 10^5$	$3.3 \pm 1.3 \times 10^4$	84.3 ± 6.4
	PCTE	$1.4 \pm 1.3 \times 10^6$	$2.2 \pm 0.6 \times 10^5$	84.0 ± 4.6
	Cellulose	$2.1 \pm 0.8 \times 10^5$	$3.5 \pm 1.9 \times 10^4$	83.4 ± 4.6
B	Cellulose	$1.4 \pm 0.7 \times 10^6$	$3.5 \pm 1.7 \times 10^5$	74.9 ± 12.4
	PCTE	$2.1 \pm 1.1 \times 10^5$	$5.0 \pm 0.5 \times 10^4$	75.8 ± 2.7
	Nylon	$2.1 \pm 1.1 \times 10^5$	$1.0 \pm 0.2 \times 10^5$	49.5 ± 9.2
	Durapore	$2.1 \pm 1.1 \times 10^5$	$5.2 \pm 1.1 \times 10^4$	74.9 ± 5.5
C	PCTE	$7.6 \pm 2.3 \times 10^5$	$1.1 \pm 0.4 \times 10^5$	85.8 ± 5.8
	Nylon	$7.6 \pm 2.3 \times 10^5$	$9.6 \pm 3.3 \times 10^4$	87.5 ± 4.3

* Average ± SD for triplicate (experiments A&B) or duplicate (experiment C) filters.
† Average of initial counts from all 1 L samples on a given experiment.

The total amount of MAA DNA isolated from the PCTE filters was significantly greater than from the nylon filters ($p<0.01$) and real-time PCR Ct values for the PCTE filtered samples were significantly lower consistent with recovery of more target DNA. (Table 4.7b). Filtrate-inoculated MGIT tubes did not consistently signal positive and those that did signal positive had a time to detection (TTD) longer than the limit of detection for the standard curve.

Table 4.7 B. Trial 2d – Recovery of DNA Using Different Pore Size Filters

Sample	Number of cells in DNA extraction	Total DNA Isolated (ng)*	PCR results (IS1311/mean Ct)
PCTE	$1.1 \pm 0.4 \times 10^5$	203.1 ± 49.5	Positive / 22.0 ± 2.1
Nylon	$9.6 \times 3.3 \times 10^4$	29.2 ± 22.2	Positive / 27.6 ± 3.1

*Average of triplicate samples.

Trial 2e investigated MAA's adherence to processing vessels as an explanation for the significant cell loss between spiking and recovery. Roughly $2 \pm 1 \times 10^2$ MAA cells were recovered from the 31 cm^2 glass base of the filter bell using a cotton swab, compared to an estimated $9 \pm 3 \times 10^5$ total MAA cells lost in filtration process.

Since the prior trial demonstrated that MAA adhered to glass vessels, Trial 2f tested a blocking agent (Sigmacote™) to try to minimize MAA cell adherence. The number of MAA cells lost was slightly higher for the treated containers than untreated filter bells ($p>0.05$): the percent of MAA cells lost with Sigmacote™-treated filter bells was 87.7 ± 9.6 % and 67.9 ± 10.4% for the untreated filter bells.

After completing these trials and selecting filtration with 0.4μm PCTE filters homogenized for inoculation to MGIT tubes as the optimal method, a finished water sample from the Pinellas Utility distribution system was tested by this protocol. A MAC species was isolated and indentified by PCR.

DISCUSSION AND CONCLUSIONS

Filtration was better than centrifugation for concentration of MAA spiked to water samples as determined both by yield of viable MAA cells and total DNA. The filter concentration technique had the added advantage of simple laboratory logistics: a single step filtration and placement of filters directly into MGIT culture media or DNA extraction vials with no additional manipulation or attendant loss of sample. The analytical sensitivity of the filter concentration technique was a direct function of the volume of water filtered.

The study demonstrated that DNA could be extracted directly from MAA cells collected on PCTE filters with no apparent inhibition of the PCR due to the presence of the filter materials in the DNA extract. The primary advantage to this method is that DNA can be extracted from all collected organisms without additional losses from attempts to elute the MAA cells from filters, dilution, or using only a portion of the collected sample. No inhibition of the PCR due to contaminants present in the Madison, WI tap water samples or disrupted PCTE filters was observed.

The loss of MAA cells during processing (~80%) with glass and plasticware was still considered unacceptably high. Hydrophobic interactions with plasticware, exaggerated by the number of plastic centrifuge bottles employed in the procedure, was a possible explanation for loss of MAA cells. Given the findings in Chapter 9 (see below), a significant number of mycobacterial cells likely remain adhered to water-containing vessels resulting in a significant underestimation of the number of these organisms in a water sample. A blocking agent such as Sigmacote™ did not improve recovery of MAA from water.

Acetone-dissolvable cellulose acetate filters will not be useful for recovering MAC cells after filtration since acetone treatment was lethal for MAA. PCTE filters were equivalent to other filter types in recovery of viable MAA cells and superior to nylon filters in recovery of MAA DNA. Also, during the ribolyser step of the DNA extraction procedure, the nylon filters did not break down as completely as did the PCTE filters, likely contributing to the lower yield of DNA.

Filtration is the preferred method of concentration of MAC bacteria from drinking water, followed by detection by real-time PCR and/or MGIT culture. If direct DNA extraction is attempted after filtration, 0.4 µm PCTE filters should be used and mechanically homogenized for culture.

CHAPTER 5
OBJECTIVE THREE

OBJECTIVE 3: ASSESS THE EFFECT OF SAMPLE DECONTAMINATION METHODS ON THE GROWTH OF MAC ISOLATES IN FINISHED WATER.

Finished water samples may contain non-target bacterial organisms that overgrow and thus mask the presence of the MAC target organisms. Decontamination methods used to suppress the growth of these non-target organisms during culture must not interfere with isolation of MAC. Sample decontamination methods initially were adopted from those used for isolation of *M. tuberculosis* from clinical samples. These methods were too harsh and caused an unacceptably high loss of MAC organisms. Since the late 1990s cetylpyridinium chloride (CPC) has been the preferred decontaminant, in spite of the findings that this detergent too causes some loss in viable MAC (Kamala et al., 1994;Neumann et al., 1997;Schulze-Roebbecke et al., 1991;Stabel, 1997;U.S.Department of Agriculture, 1974;Whipple and Merkal, 1983). Falkinham recognized that "the best approach would be to avoid decontamination" and suggested that selective concentration of mycobacteria from water using immunomagnetic beads would be a preferred approach (Falkinham, III, 2002).

The MAC quantification methods developed in this study were used to determine baseline MAC susceptibility to chlorine and CPC. Given the increased sensitivity of the isolation and quantification methods developed in this project, a re-examination of decontaminants routinely used in water distribution systems was needed. After this re-examination, rather than exploring the almost endless list of possible other decontaminants, the project then moved to immunomagnetic bead development (Objectives 4 & 5) since results from this MAC organism concentration technique would affect the choice of optimal decontamination protocols.

METHODS AND MATERIALS

Chlorine. The sensitivity of MAC bacteria to chlorine was evaluated using test strains MAC ATCC 35712 and MAP ATCC19698. Log-phase cells were exposed to 0, 1, 2 and 3 mg/L free chlorine (as NaOCl). The chlorine-exposed cells were incubated at room temperature and sampled at 10, 30 60 and 120 minutes. At each sample point, free residual chlorine was measured and, after neutralization of chlorine using sodium thiosulfate, viable MAA and MAP counts were determined by the MGIT counting method established in Objective #1. For a subset of samples, agar plates were inoculated to determine the number of colony forming units (CFU) as a means of corroborating the CFU estimates by MGIT methods. Cellular ATP was also measured as an indicator of cell viability using the BacTiter-Glo™ assay (Promega, Madison, WI).

Cetylpyridinium chloride (CPC). The sensitivity of MAA strain ATCC 35712 to CPC was tested. Log-phase and lag-phase (stationary growth phase) cells in 1L of sterile water were exposed to 0.05% and 0.005% CPC by weight in a 1L glass bottle. The experiment included a control with no CPC. Prior to introduction of CPC, a sample of the MAA suspension in each glass bottle (control, 0.005% and 0.05%) was placed in MGIT media in triplicate for enumeration. Upon introduction of CPC, the glass bottle was shaken for 30 seconds to thoroughly mix CPC with water. MAA cells were incubated in CPC at room temperature for 30 minutes. After incubation, the bottle was shaken again, cells were filtered and the filter washed with sterile water to remove the CPC. The filter was placed in a MGIT tube for enumeration.

For the 0.005% CPC log-phase trial, samples of the MAA suspension after shaking were put directly to MGIT culture tubes in triplicate without filtration.

RESULTS

Chlorine. There was a linear dose-response relationship to chlorine concentration (response being cell death) for both MAA (r^2 = 0.95) and MAP (r^2 = 0.85) using a constant exposure time of 120 minutes (Figure 5.1), although in no case did complete killing occur. The effect of chlorine on MAA viability was found to be the same regardless of cell counting method (Figure 5.2). For MAA, viable counts were significantly reduced by exposure to 2 mg/L NaOCl for ≥30 minutes or 3 mg/L NaOCl for ≥10 minutes (Figure 5.3). For MAP, viable counts were significantly reduced by exposure to 3 mg/L NaOCl for ≥10 minutes (Figure 5.4). The ATP content of MAP cells was significantly decreased ($p<0.01$) with exposure to 1, 2 and 3 mg/L chlorine as compared to untreated cells (data not shown).

CPC. In two trials, exposure to the 0.005% CPC solution for 30 minutes did not affect the number of cells enumerated from inocula placed directly into culture tubes (unfiltered). That is, no differences in MAA counts were seen between the no-CPC controls and CPC-exposed cells when pipetted directly into culture tubes without filtration (Figure 5.5).

In Figure 5.5, note the reduced recovery after filtration shown in the four control samples not exposed to CPC: at least a log of cells disappears from each. This effect appears amplified for the test samples exposed to CPC at both concentrations, that is, more cells are lost than were lost from filtration alone of the no-CPC controls.

DISCUSSION AND CONCLUSIONS

As expected, both MAA and MAP were relatively resistant to killing by free chlorine. Concentrations of 3-4 mg NaOCl/L for 120 minutes did not result in complete killing of the MAC test strains. MAP was more chlorine-resistant than MAA. This finding is consistent with other published studies and explains the persistence of MAC in water distribution systems (Whan et al., 2006;Whan et al., 2001). It may be possible to capitalize on the chlorine resistance of MAC for water sample decontamination procedures since other contaminants are susceptible to these levels of chlorine (Norton et al., 2004).

The CPC trials again underscored the significant loss of cells that occur in sample processing since no loss was seen in unprocessed (direct inoculation) samples. When the CPC was added to the cell/water solution and the bottle shaken, significant foaming occurred. CPC may have surfactant properties which would affect the surface attachment characteristics of both the container and the organisms. This issue is investigated further as part of Objective 7, Chapter 9.

These findings add to the body of water decontamination literature in associating common decontamination methods with a new method for counting mycobacterial contaminants in finished water: (1) use of free chlorine at standard water distribution system concentrations to minimize contaminating non-mycobacterial organisms is unlikely to interfere with isolation of MAC species and (2) While MAC are resistant to CPC, it appears that this frequently used decontaminant may augment hydrophobic partitioning of mycobacteria and interfere with accurate MAC quantification.

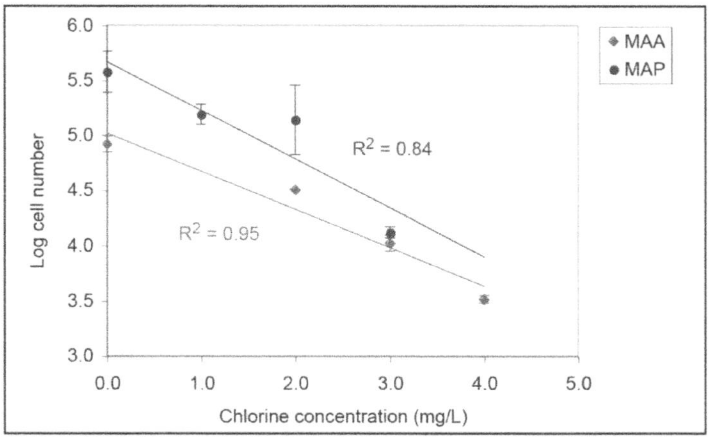

Figure 5.1 Effect on Log Phase MAA and MAP of Free Chlorine Exposure for 120 Min

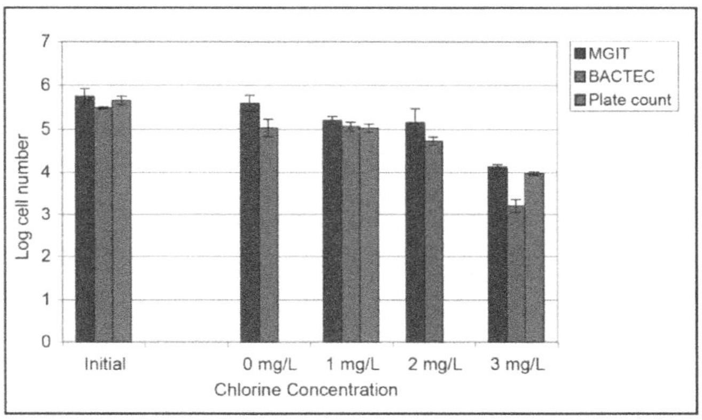

Figure 5.2. Impact of Cell Counting Method on Evaluating Free Chlorine's Effect on MAA Viability

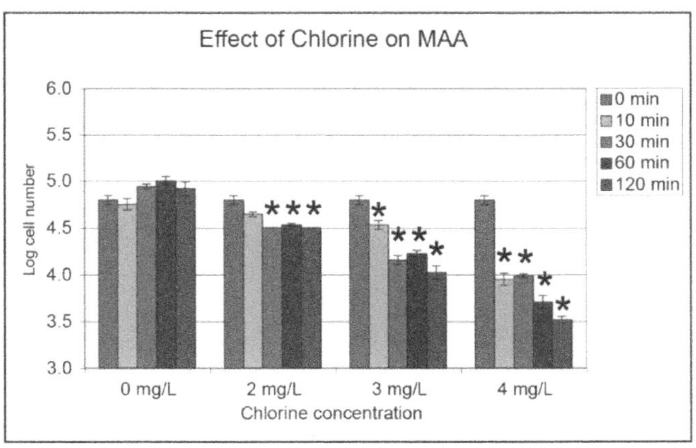

Figure 5.3. Effect of Differing Concentrations of Free Chlorine on MAA.

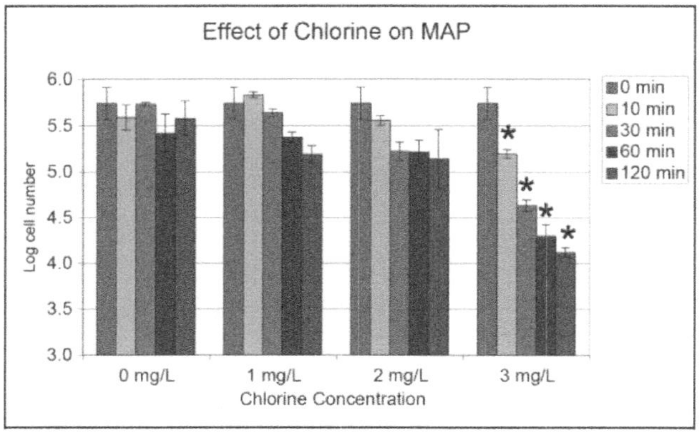

Figure 5.4. Effect of Free Chlorine on MAP.

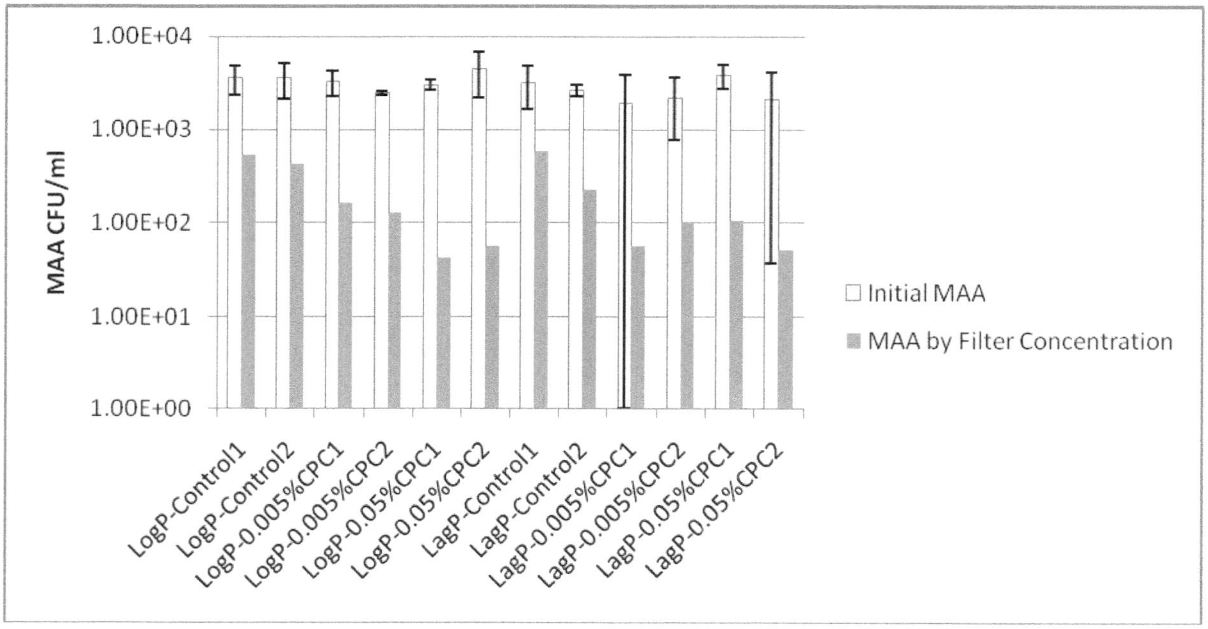

Figure 5.5. Recovery of Filter-Concentrated Log and Lag-Phase MAA After Exposure to 0.05% and 0.005% CPC (95% Confidence Intervals Shown).

CHAPTER 6
OBJECTIVE FOUR

OBJECTIVE 4: EVALUATION OF ANTI-MAC ANTIBODIES FOR IMMUNOMAGENTIC SEPARATION (IMS).

This objective focused on identifying and characterizing a MAC-specific antibody for immunomagnetic organism capture techniques with beads. Since the number of MAC organisms is likely to be very low in finished water, capturing and concentrating these few organisms in a small volume is expected to increase analytical sensitivity whether by culture, PCR or a combination of both methods.

METHODS AND MATERIALS

Antibodies. A number of commercially-available and experimental monoclonal and polyclonal antibodies were evaluated for reactivity to MAC organisms. (Table 6.1.)

Table 6.1. Summary of Antibodies Tested For Reactivity to MAC Species

Antibody (species)	**Supplier**	**Reported reactivity**
mAb 18101 (mouse)	QED, Inc.	MAA
mAb 18102 (mouse)	QED, Inc.	MAA
mAb 18103 (mouse)	QED, Inc.	MAA
mAb 18104 (mouse)	QED, Inc.	MAA
mAb BM1197 (mouse)	Ascris, Inc.	MAA
mAb FDA "D" (mouse)	US FDA	MAA
mAb FDA "E" (mouse)	US FDA	MAA
mAb CS35 (mouse)	CO State University	*M. leprae* LAM
pAb Aq#6 (rabbit)	JTC	MAP
pAb DAKO (rabbit)	DAKO, Inc.	*M. bovis*
pAb Grant (rabbit)	Irene Grant, Belfast Ireland	MAP
pAb Biodesign (rabbit)	Biodesign International	*M. tuberculosis*
pAb1 CSU (rabbit)	CO State University	*M. leprae* surface antigen minus LAM
pAb2 CSU (rabbit)	CO State University	*M. leprae* antigen 85
pAb3 CSU (rabbit)	CO State University	*M. leprae* cell wall minus LAM
pAb Yvette (cow)	JTC	Naturally infected cow
pAb Snyder (cow)	JTC	Naturally infected cow
pAb Rbt anti paratb (rabbit)	JTC	MAP
pAb Rbt anti MAC (rabbit)	JTC	MAA
pAb Chx anti-paratb (chicken)	JTC	MAP
pAb Chx anti-MAC (chicken)	JTC	MAA

Antibody labeling methods. Antibodies were tested using a range of labeling conditions including variations in antibody dilution, incubation time and temperature, washing conditions, and initial cell concentration. A summary of the general labeling approach follows is found in Appendix B.

Organisms used to evaluate cross-reactivity. To assess an antibody's specificity for MAC, antibody preparations were tested against a variety of non-MAC mycobacteria and other bacterial species. Reactivity was measured by flow cytometry. A list of test organisms is provided below in Table 6.2. Several non-mycobacterial isolates were tested for reactivity to anti-mycobacteria antibodies. These included control strains used for testing in the WSLH Water Microbiology laboratory as well as stock culture organisms used in the WSLH Bioaerosol/Mycology laboratory.

Table 6.2 List of Organisms Tested (Number of Isolates)

MAC	Non-MAC mycobacteria	Other organisms
M. avium (9)	*M. asiaticum*	*Aeromonas hydrophila*
M. avium ssp. *intracellularae* (4)	*M. gordonae*	*Agrobacter globiformis*
M. avium ssp. *paratuberculosis* (3)	*M. fortuitum (2)*	*Citrobacter freundii*
	M. lentiflavum (2)	*Enterobacter aerogenes*
	M. malmoense (2)	*Cryptosporidium parvum*
	M. marinum (2)	*Encephalitozoon intestinalis*
	M. scrofolaceum	*Enterobacter cloacae*
	M. silvaticum	*Enterococcus faecalis*
	M. simiae/simiae complex (2)	*Escherichia coli(8)*
	M. terrae complex (2)	*Giardia lamblia*
		Klebsiella pneumoniae
		Nocardia asteroids
		Nocardia facinia
		Nocardiopsis dassonvillei
		Proteus vulgaris
		Pseudomonas aeruginosa
		Rhodococcus equii
		Sphingomonas paucimobilis
		Stenotrophomonas maltophila
		Streptomyces rimosus

Water microbiology stocks (*Aeromonas hydrophila* ATCC 7966, *Enterobacter aerogenes* ATCC 13048, *Enterococcus faecalis* ATCC 19433, *E. coli* ATCC 11229, 11775, 25922, *Klebsiella pneumoniae* ATCC 13883, *Proteus vulgaris* ATCC 13315, *Pseudomonas aeruginosa* ATCC 10145) were inoculated from nutrient agar slants into 1 mL PBS and the organism concentration enumerated using flow cytometry as described above. Frozen stocks of *Agrobacter globiformis, Citrobacter freundii, Enterobacter cloacae, Nocardia asteroides, Nocardia facinia,Nocardiopsis dassonvillei, Rhodococcus equii, Sphingomonas paucimobilis, Stenotrophomonas maltophila,Streptomyces rimosus* were obtained from the WSLH Bioaerosol/Mycology laboratory, thawed at room temperature and streaked onto either trypticase soy agar or Middlebrook 7H9 agar. Plates were cultured for 24-72 hours and the cells inoculated into 1 mL PBS before determining the concentration of cells in the suspensions.

Other non mycobacteria organisms tested for cross reactivity included *Cryptosporidium parvum* IOWA isolate (Sterling Parasitology), *Encephalitozoon intestinalis (*ATCC 50603, Sterling Parasitology), *Giardia lamblia (H3 isolate, Waterborne Inc.)* and *Escherichia coli* WSLH environmental isolates 1-5 (isolated from natural water samples)

Source and culture of mycobacteria isolates. The MAC and non-MAC mycobacteria tested throughout the course of this project were obtained from a variety of sources. The JTC and the US EPA provided several clinical (JTC) and environmental (EPA) MAC isolates. The remaining non-MAC mycobacteria were provided by the WSLH's Clinical Disease Division. Each of these isolates was cultivated on Middlebrook 7H9 slants or in Middlebrook 7H9 broth.

Antibody absorption. To improve the specificity of the polyclonal IgY antibodies, cross-reactive antibodies from antisera were absorbed with *M. phlei* or *E. coli* whole cells (cross-reacting antibodies bind to the antigens presented by the non-MAC cells and are thus removed). Briefly, 50 to 1000 µg of antibody were absorbed with 10^5 to 10^8 CFU of either *M. phlei* ATCC 11758 or *E. coli* DH 5. After a one-hour incubation at 37°C, the mixture was filtered through a 0.2 µm pore-size membrane filter to remove the bacteria. The specificity of the absorbed antibodies was independently assessed by flow cytometry as described previously.

Antibody avidity. MAC IgY antibody was evaluated for avidity (degree of binding) using flow cytometric methods developed and published by the research team for use with *Cryptosporidium* and *Giardia* antibodies (Hoffman et al., 1999). Cells were labeled with the candidate antibody as described above, the excess primary antibody was removed by washing with 1 mL reagent grade water and the antibody-organism complex was resuspended in an 8M urea solution for 3 minutes at room temperature. Following incubation, the urea was removed by washing 3 times in PBS and the cells were labeled with the secondary antibody as described above. Cells were analyzed by flow cytometry and the fluorescence of the treated cells was compared to that of untreated controls. The experimental differences in fluorescence of the treated and untreated samples were evaluated using a student's t-test.

Flow cytometry. Samples were analyzed by flow cytometry (EPICS XL, Beckman Coulter, Fullerton, CA) and events were plotted on histograms depicting fluorescence on the x-axis and cell count on the y-axis. Fluorescence of the antibody-labeled organisms was compared to that of non-target organisms and secondary-only controls to determine the extent of specific fluorescence. Antibodies showing reactivity to MAC were screened further against a collection of mycobacterial and non-mycobacterial isolates.

RESULTS

Antibody reactivity and specificity. A summary of each antibody's performance in binding to MAC species is listed in Table 6.3. None of the 8 monoclonal antibodies evaluated showed the level of sensitivity needed for IMS development. The QED monoclonal antibodies (mAbs 18101-18104) and the Ascris antibody (mAb BM1197) did not label any of the isolates tested. The other three monoclonal antibodies obtained from the FDA (mAbs D and E) and the Colorado State University monoclonal (CS37) labeled some MAC isolates weakly or labeled a subpopulation of the cells only. All of the 13 polyclonal antibodies evaluated showed some degree of reactivity with at least one mycobacterial isolate. However, the majority of those tested also cross-reacted with *E. coli*.

Most promising among the antibodies were those produced by the JTC: the polyclonal antibodies produced in rabbits in response to inoculation with MAA culture filtrate and polyclonal antibodies produced in chickens inoculated with the MAA cells. The least amount of cross reactivity to *E. coli* or other mycobacteria was seen with the chicken anti-MAA antibody. Therefore this antibody was chosen for further evaluation.

Table 6.3. Antibody Vs. Organism Binding Results

Antibody	Organisms (No. of isolates) tested	Results
mAb 18101	MAA; MAP (2); *M. intracellularae*	Did not label.
mAb 18102	MAA; MAP (2); *M. intracellularae*	Did not label.
mAb 18103	MAA; MAP (2); *M. intracellularae*	Did not label.
mAb 18104	MAA; MAP (2); *M. intracellularae*	Did not label.
mAb BM1197	MAA; MAP (2); *M. intracellularae*	Did not label.
mAb FDA "D"	MAA; MAP (2); *M. intracellularae*	Weakly labeled *M. intracellularae* and MAP clinical isolate. Did not label MAA or ATCC MAP.
mAb FDA "E"	MAA; MAP (2); *M. intracellularae*	Weakly labeled M. intracellularae and MAP clinical isolate. Did not label MAA or ATCC MAP.
mAb CS35	MAA (4); MAP (2), *M. intracellularae; M. asiaticum; E. coli (2); K. pneumoniae*	Antibody labels a subpopulation of mycobacteria in suspension intensely. Expression of the antigen is not uniform in mycobacteria cultures.
pAb Aq#6	MAA (7); MAP (2); *M. intracellularae (3); M. asiaticum; M. malmoense; E. aerogenes; E. coli; K. pneumoniae*; aged MAA and *M. ptb.*	Antibody labeled nearly all of the mycobacteria isolates tested although not as intensely as the DAKO Ab below. Cross reacts with E. coli.
pAb DAKO	*M avium (7); MAP (4); M. intracellularae (4); M. asiaticum ; M. fortuitum complex (2); M. malmoense; M. terrae complex (2); M. simiae complex (2); M. lentiflavum (2); M marinum (2). E. coli (7); E. faecalis; K. pneumoniae; A. hydrophilia; P. aeruginosa; P. vulgaris.*	Antibody labeled nearly all of the mycobacteria isolates tested. Cross reacts with E. coli.
pAb Grant	*MAA, MAP, M. intracellularae; E. coli (2)*	Weak reactivity with mycobacteria isolates at high titer. Side by side with DAKO and Aq#6 showed it least reactive of the 3 pAbs test. Cross reacts with E. coli.
pAb Biodesign	MAA, MAP; *M. intracellularae, E. Coli*	Weakly labeled MAA. Did not label M. ptb. or M. intracellularae. Cross reacts with E. coli.
pAb1 CSU	MAA (4); MAP; M. *simiae*; *M marinum*; *M. intracellularae, E. Coli*	Moderate to good labeling of mycobacteria. Low to moderate cross reactivity with E. coli
pAb2 CSU	MAA; MAP; *E. coli*	Dim labeling of MAA, no labeling of M. ptb.
pAb3 CSU	MAA; MAP; *E. coli*	Good labeling of mycobacteria. Cross reacts with E. coli.
pAb JTC "Eyvete "	MAA; MAP; *E. coli*	Good labeling of mycobacteria. Strongly cross reacts with E. coli.
pAb JTC "Snyder "	MAA; MAP; *E. coli*	Good labeling of mycobacteria. Strongly cross reacts with E. coli.
pAb JTC rabbit anti-MAP	MAA; MAP; *E. coli*	Intense labeling of mycobacteria. Cross reacts with E. coli.
pAb JTC rabbit anti-MAC	MAA; MAP; *E. coli*	Intense labeling of MAA. Good labeling of M.paraTB. Cross reacts with E. coli.
pAb JTC chicken anti-MAP	MAA; MAP; *E. coli*	Good labeling of mycobacteria. Cross reacts with E. coli.
pAb JTC chicken anti-MAC	MAA; MAP; *E. coli*	Good labeling of mycobacteria. Minimal reactivity with E. coli.

Absorption results: To further reduce the nonspecific reactivity of the chicken anti-MAC IgY (heretofore MAC IgY), it was absorbed with *E. coli* and/or *M. phlei* as described above. The antibody was rescreened for reactivity to MAC and also evaluated for reactivity to *E. coli* using flow cytometric methods. Reactivity with MAC was retained while the cross-reactivity with non-target *E. coli* was greatly diminished (Table 6.4).

Table 6.4. Effect on Fluorescence Before and After Absorption*

Organism	Log MF before absorption	Log MF after absorption
MAA ATCC 35712	8.15	9.38
E. coli env. isolate 2	6.3	0.143
E. coli env. isolate 5	3.09	0.147
E. coli ATCC 11229	38.2	0.148

*Removal of cross-reacting antibody by exposure to *E.coli* and *M. phlei*

Antibody avidity: The strength of the antibody-antigen complex was evaluated using avidity determination methods described above. The log mean fluorescence intensity of the treated cells (mean 6.57, SD 0.28) labeled with MAC IgY was slightly lower than that of the untreated controls (mean 7.82, SD 1.48), however, the results were not statistically different ($p=0.13$). These data indicated that the MAC IgY was a good candidate for IMS bead development.

Reactivity of anti-MAC IgY: The mean fluorescence (MF) values of the MAC, non-MAC mycobacteria and non-mycobacteria isolates are summarized in Table 6.5. Although the values were not statistically different, the mean fluorescent values of MAC exceed non-MAC mycobacteria which exceed fluorescence values for non-mycobacteria. These data suggest that the anti-MAC IgY is specific to MAC, particularly MAC sub-species MAA and MAP.

Table 6.5. Summary of Log Mean Fluorescence (MF) Values of Organisms When Labeled With Anti-MAC Igy

Organism group	Log MF	SD	Range
All MAC	4.71	2.13	1.52-9.68
MAA/MAP	5.42	1.89	2.90-9.68
Environmental MAA (EPA)	5.39	0.72	4.63-6.05
Clinical MAA (WSLH)	3.94	1.06	2.90-5.01
Mycobacteria (non MAC)	1.64	0.38	1.22-2.30
Non mycobacteria	1.61	0.96	0.69-3.49

Reactivity of anti-MAA IgG: The purified and absorbed polyclonal rabbit anti-MAA IgG cross reacted with all mycobacteria tested. It reacted slightly with the non-mycobacterial isolates *E. coli* ATCC 25922 and *Proteus vulgaris* ATCC 13315 (data not shown). Although less specific than the anti-MAC IgY, the research team chose to send the rabbit antibody to Invitrogen for IMS bead development to maximize analytical sensitivity. Molecular identification methods (PCR) would be relied upon to further speciate any organisms that would be non-specifically bound by the IMS beads.

DISCUSSION AND CONCLUSIONS

Many of the polyclonal antibodies evaluated for this project cross-reacted with *E. coli* and other unrelated bacteria. Much of the initial antibody work performed in this project used an antibody produced by the DAKO Corporation. Unfortunately, production and sale of this antibody was discontinued during the course of project causing a loss of time and effort and forcing continued search for a suitable antibody. Fortunately, the anti-MAC IgY and the anti-MAP IgG kindly donated by the JTC were both sufficiently specific and capable of binding to MAC cells with high avidity. They were good candidates for IMS bead production in Objective 5.

CHAPTER 7
OBJECTIVE FIVE

OBJECTIVE 5: EVALUATION OF IMS BEAD CONSTRUCTS FOR SELECTIVE MAC CONCENTRATION.

METHODS AND MATERIALS

Evaluation of IMS bead products: Can didate antibodies were conjugated by Invitrogen (formerly Dynal, Frederick MD) using proprietary methods. Upon receipt of the IMS beads the binding of MAA (ATCC 35712) cells was assessed using a range of concentration, temperature and buffer conditions. Limited trials were also conducted using MAP (ATCC 19698). The general approach follows: SYTO-9 stained cells were suspended in 1 mL PBS with 0.01% Tween 20 and 0.1-1.0% BSA. Ten to one-hundred µL IMS beads were added and the tubes rocked at either room temperature or at 37ºC for 20 – 60 minutes. Tubes were placed in the magnetic particle concentrator (MPC) and rocked gently for 2 minutes. The tubes were opened and the supernatant gently aspirated so as not to disturb the bead pellet on the side of the tube. The supernatant was transferred to a labeled 12 x 75 mm tube. The beads were then washed 1-2 times in 1 mL PBS or reagent water with 0.01% Tween 20 and each time the wash buffer was pooled with the initial supernatant. The number of mycobacteria in the pooled supernatant-wash buffer suspensions was enumerated using flow cytometry and FlowCount beads as described in Chapter 3 (Objective 1). In these IMS experiments, however, the addition of particulates (beads and bead impurities) necessitated quantification of organisms based on fluorescence (SYTO-9) and light scatter rather than on light scatter properties alone. In order to account for cell loss associated with adherence to plasticware, the seeded dose (or initial dose) was determined using identically processed "no bead" controls and the following equations was used to express the results:

$$\frac{\text{(Cells in "no bead" control supernatant} - \text{cells in test sample supernatant)}}{\text{Cells in "no bead" control supernatant}} \times 100 = \%\ \text{capture}$$

Tosyl-activated IMS bead development and evaluation: DynaBeads® M280 Tosylactivated are superparamagnetic beads that have been coated with polyurethane. The hydroxyl groups on these beads have been activated to covalently bind the amino or sulfhydryl group of potential ligands. Both the rabbit anti-MAC IgG and the chicken anti-MAC IgY were coated onto Invitrogen's tosyl-activated IMS beads according to the package insert. Briefly, 0.5 mL of a 100 µg/mL solution of each antibody were incubated with tosyl-activated beads that had been washed in borate buffer and resuspended in 0.5 mL borate buffer. Tubes were rocked overnight at 35ºC and the IMS beads were washed 4 times in BSA containing PBS and TRIS as directed in the package insert. Capture of MAA was assessed according to the protocol described above. Limited trials were also conducted using MAP ATCC 19698 as the target organism.

Biotinylation: The anti-MAC IgY antibody was biotinylated using a EZ Link SulfoNHS Biotinylation kit (Pierce Biotechnology, Rockford IL) according to the package insert. Biotin incorporation was verified spectrophotometrically and antibody functionality was confirmed by labeling mycobacteria with biotinylated anti-MAC IgY antibody followed by streptavidin-FITC.

A [biotin anti-MAC IgY] – [streptavidin bead] construct was prepared by incubating streptavidin coated IMS beads with the biotinylated antibody. The antibody-biotin-streptavidin-bead construct was washed 4 times with 1 mL PBS supplemented with 0.1% BSA and 0.01% Tween 80. Beads were tested for the ability to capture MAA using the IMS protocol described above. The functionality of the biotinylated anti-MAC IgY was also assessed by labeling MAA with biotinylated antibody, washing 3 times with 1 mL reagent grade water supplemented with 0.01% Tween 20. IMS using Invitrogen's streptavidin coated beads was performed in the manner described above.

RESULTS

Two polyclonal antibodies (chicken anti-MAC IgY and rabbit anti-MAP IgG) were sent to Invitrogen for IMS bead production. In addition, IMS bead construction was attempted at the WSLH using tosyl-activated and streptavidin coated beads. A summary of the results of the various IMS experiments are presented in Table 7.1. The efficiency of capture for all IMS products was low (less than 44%). Modifications of buffer composition, time and temperature, did not enhance recovery (data not shown). With the exception of the IgG–MyOne construct, no mycobacteria were visualized when beads were applied to slides and examined under epifluorescent microscopy. Repeat of the IgG-MyOne Ig experiment using identical conditions showed no visible cells suggesting that cells may have been physically entrapped by the magnetic particles rather than captured via antibody interaction.

Table 7.1 Capture of MAA ATCC 35712 by IMS Bead Constructs

Bead construct	Recovery range (%)	Number of trials
Ig Y – M280-1	0 - 5.8	4
Ig Y – M280-2	0 – 8.0	3
Ig G – M280	0 – 44	6
Ig Y biotin – streptavidin bead	3.7 - 26.7	3
Ig Y – tosyl activated bead	1.3	1
Ig G – tosyl activated bead	0.0	1
Ig Y – MyOne bead	0.0	2
Ig G – MyOne bead	0 – 18.7	4

The development of an IMS product capable of efficient capture of MAA with either of these purified and absorbed antibodies was not successful. One explanation is that the antibodies were not conjugated in a proper or functional orientation on the IMS bead. The presence of the antibody on the beads was verified at WSLH by labeling immunoglobulin-coated beads with fluorescein conjugated anti-immunoglobulin and verifying fluorescence and also at Invitrogen using a protein detection assay. Neither the research team nor the team's Invitrogen collaborators however could demonstrate specific reactivity with any of the constructs produced.

Another possible explanation for lack of binding relates to the hydrophobicity of the mycobacterial cell wall. It is possible that mycobacteria were repelled by the hydrophilic nature of the IMS beads. Both the M280 and the MyOne beads are hydrophilic but despite continual mixing and lengthened incubation periods the antibody conjugated bead was not capable of binding MAA or in limited experiments MAP ATCC 19698. The tosyl-activated beads are hydrophobic, however, attachment of mycobacteria to these beads was not seen.

DISCUSSION AND CONCLUSIONS

Commercially available antibodies making claims to bind MAC bacteria and even function on IMS beads for immunoconcentration from commercial and private sources failed to show sufficient MAC binding avidity and specificity by flow cytometry to warrant forwarding to our commercial partner Invitrogen for coupling to magnetic beads. However, a novel anti-MAC IgY antibody and rabbit anti-MAC IgG preparation created the JTC, did show promising flow cytometry results. Unfortunately, these antibodies were not suited for immunomagnetic concentration of MAC from water using beads.

CHAPTER 8
OBJECTIVE SIX

OBJECTIVE 6: ESTABLISH AND EVALUATE REAL TIME CAPACITY FOR A VALIDATED MULTIPLEX PCR TO DETECT AND QUANTIFY MAC DNA ISOLATED FROM WATER SAMPLES.

PCR technology offers useful tools for detecting, identifying and quantifying bacterial organisms. These assays offer not only detection of the presence of the organism, but offer finer discrimination among mycobacterial species within the same complex than can be accomplished with other assays such as high performance liquid chromatography.

Numerous molecular typing methods have been used to try to distinguish among the MAA. Genetic insertion sequences (IS) that vary in type, number and location on the genome are useful targets for species identification. These sequences are characterized by a small degree of mobility and limited polymorphism. The insertion element IS*900* is consistently found in MAP and is present in 15-20 copies. There are 2-13 copies of the insertion element IS*901* in virulent bird-type *M. avium* ss. *avium* (*MAA*) but this element is not found in most environmental or human isolates of *MAA*. *M. avium* and *M. intracellulare* contain the insertion element IS*1245* but MAP does not. Its copy number is highly variable depending on the strain serotype. A closely related IS element, IS*1311* (85% sequence identity with IS*1245* at the DNA level), is found in all *MAC* species including MAP.

This project developed two multiplex (i.e., multiple target) single reaction PCRs by selecting and evaluating primers for MAC target insertion sequences and optimizing PCR conditions for extracting DNA from drinking water matrices. One assay (PCR A) employed the MultiCode RTx system developed by EraGen Biosciences for real-time (quantitative) PCR and the other was a new generation of a multiplex assay previously developed at the Johne's Testing Center (PCR B). (Experiment details are found in Appendix C.)

PCR A: ERAGEN RT-PCR

Each major component of this assay was assessed independently, then combined to build the final RT-PCR.

Evaluation of IS1311, IS900 and IS1245 primers. Primers were designed by EraGen to amplify IS in MAC organisms for the MultiCode RTx system. Two primer sets were designed for each IS target (designated 1 and 2). PCR conditions were optimized, specificity was evaluated and the final designs were selected for primers targeting IS*1311* (present in MAA and MAP and some *M. intracellulare* strains), IS*900* (present only in MAP) and IS*1245* (present primarily in MAA and in some strains of MAP).

Annealing temperature. Initially, an annealing temperature of 65°C was selected for the PCR cycling protocol with these primers. At this annealing temperature, good specificity was seen (MAA DNA amplified and none of the non-target DNA amplified within 45 PCR cycles). However, the limit of detection for MAC organisms was relatively high and the standard Internal Amplification Control (IAC) for the MultiCode RTx system did not amplify. In addition, the IS*1311* primers did not amplify any of the tested *M. intracellulare* strains tested, despite reports in the literature showing IS*1311* elements present in some *M. intracellulare* strains (Keller et al., 2002).

The specificity of the primers was evaluated at lower annealing temperatures (Table 8.1) the IS*1311*-2 primers at an annealing temperature of 62°C amplified several of the non-target bacterial DNA. When the annealing temperature for these primers was increased to 63°C several of the DNA samples that amplified at 62°C did not amplify, e.g. *M. intracellulare*, *M. phlei*, *M. lentiflavum*, *N. asteroids* and *S. griseus*. The IS*900*-2 primers showed similar results to the IS*1311*-2 primers i.e. several non-target strains amplified at 62°C. However, with both of these primer sets, the Ct for the non-target DNA tended to be over 40 cycles, whereas 50 ng of target DNA amplified with a Ct between 17 and 20 cycles. Lowering the annealing temperature to either 62°C or 63°C significantly decreased the limit of detection for MAC DNA.

Table 8.1. Specificity of IS*1311*-1 and IS*900*-2 Primers with 63°C and 65°C Annealing (Melting) Temperatures for MAA (12 Strains), MAP (Strains), Non-MAC Mycobacteria (9 Species, 14 Strains) and Non-Mycobacteria (8 Species, 15 Strains)

Target organisms	IS*1311-1* Ct	*65°* Mt	IS*1311-1* Ct	*63°* Mt	IS*900-2* Ct	*65°* Mt	IS*900-2* Ct	*63°* Mt
MAA (12)	23.7± 2.2	87.2 ± 0.16	21.8±2.8	87.6±0.2	n.a.	n.a.	n.a.	n.a.
MAP (6)	26.1±8.9	87.2±0.19	21.6±6.7	87.6±0.2	22.8±2.8	86.6±0.1	22.2±2.9	86.7±0.2
MAI (5)	n.a.	n.a.	n.a.	n.a.	n.a.	n.a.	n.a.	n.a.
Non-MAC	n.a.	n.a.	n.a.	n.a.	n.a.	n.a.	n.a.	n.a.
Non-myco	n.a.	n.a.	n.a.	n.a.	n.a.	n.a.	n.a.	n.a.

Ct = mean crossing threshold. Mt = mean melting temperature. n.a. = no significant amplification.

All of the MAA and MAP target DNA samples tested amplified as expected for both of the primer sets. The MAI DNA did not amplify with these primers and the non-target DNA samples did not amplify at a level beyond assay background. Based on these data, a 63°C annealing temperature should be used for both the IS*1311* and IS*900* primers to optimize specificity.

Targeting IS1245. In addition to the primer to the IS*1311* and IS*900* elements, primers to IS*1245* were tested. The IS*1245*-2 primers amplified DNA only from the MAA strains and not from the MAP or *M. intracellulare* strains, as expected from the host range of this element as described in the literature (Johansen et al., 2005)(Guerrero et al., 1995 and Johansen, et al., 2005). There have been some reports in the literature reporting IS*1245* elements (or highly homologous elements) in some strains of *M. intracellulare* (Beggs et al., 2000;Keller et al., 2002). However, primers designed to this element did not amplify any of the *M. intracellulare* strains tested. Since the target range of the IS*1245* primers was redundant with the IS*1311* primers, they were not included in the final assay.

Addition of Internal Amplification Control. EraGen BioSciences had developed several Internal Amplification Control (IAC) DNA template and primer sets that were available for their MultiCode RTx system. However, none of these amplified in reactions with an annealing temperature of 65°C. EraGen was able to design two primer sets (designated EB108 and EB110) that amplified their IAC DNA in a temperature range from 62-65°C. Both of these primer sets were compatible with IS*1311* and IS*900* primer sets without interference of amplification or dimer formation (experiments done by researchers at EraGen, data not shown). The IAC primers sets were labeled with a HEX fluorescent marker. The IS*1311* and IS*900* primers were labeled with fluorescein (FAM), thus allowing detection of the two different amplification products on two separate color channels.

Ribolyser DNA extraction. A method for DNA extraction from MAC cells using a bead-beating (ribolyser) method resulted in good lysis efficiency and high yields of DNA from both MAA and MAP strains (details in Appendix 3). This method worked well for MAC from culture and concentrated water samples spiked with target cells and collected by filtration or centrifugation. Purification of the DNA by phenol extraction resulted in minimal loss of DNA and resulting samples were clean enough for amplification with the triplex PCR method with no inhibition of IAC amplification.

Biplex assay. After the development of the Internal Amplification Control (IAC), the DNA samples extracted from the serially diluted MAA cell suspensions were used in a PCR containing both the IS*1311*-2 and IAC primers. In this biplex reaction, the Ct for the IS*1311*-2 amplification was very similar to the Ct observed in reaction with the IS*1311*-2 primers alone. These results indicated that the presence of the IAC DNA or primers did not interfere with amplification of the IS*1311* target. The IAC amplified with an average Ct of 36.8 ± 0.9 in the presence of the MAA DNA (Table 8.2). The average Ct for the IAC for 4 water-only controls (no template DNA) was 37.2 ± 1.1 cycles (data not shown), which was not significantly different ($p<0.01$) than reactions containing MAA DNA. These results indicated the presence of the MAA DNA did not interfere with the amplification of the IAC and nothing in the DNA extraction inhibited PCR efficiency.

Limit of detection: MAA cell suspensions. To establish the limit of detection for DNA extraction and PCR methods, 5 mL of log-phase MAA ATCC 35712 cultures were centrifuged, washed, and resuspended in 5 mL PBS buffer. The cell suspensions were serially diluted down to 10^{-5}. From each of the dilutions, 100 µL aliquots were used to inoculate MGIT tubes in duplicate to determine the number of cells in each dilution. Additional 100 µL aliquots were transferred to ribolyser tubes and used in the DNA extraction procedure. The amount of DNA in each sample was quantified using the Quant-it High Sensitivity Assay Kit® (Invitrogen). Real-time PCR assays were done using 5 µL of DNA extract and standard PCR conditions. (Initial PCR experiments were completed prior to the development of the internal amplification control).

The limits of detection for the MAA DNA extractions with the PCR method was approximately 10^3 cells / 100 µL extraction volume using the IS*1311* primers and a 63°C annealing temperature (Table 8.2). In separate experiments, the limit of detection for the PCR at 62°C and 63°C was shown to be the same but to increase to approximately 10^5 cells / extraction when the annealing temperature was raised to 65°C (data not shown). These results support the decision to use an annealing temperature of 63°C for the IS*1311* primers.

Table 8.2 PCR Amplification of IS*1311* Target from DNA Extracted from Serially Diluted MAA ATCC 35712

Sample dilution	Number of cells (per MGIT)	DNA detected per reaction (ng)*	Crossing threshold (cycles) at 63°
Undiluted	Above limit	35.0	18.7
10^{-1}	3.31×10^{5}	0.9	27.7
10^{-2}	4.25×10^{4}	~0.1	34.7
10^{-3}	1.48×10^{3}	~0.01	39.2
10^{-4}	6.35×10^{1}	~0.001	n.a.
10^{-5}	1.26×10^{1}	~0.0001	n.a.
External blank	n.a.	n.a.	n.d.
Water control	n.a.	n.a.	n.a.

*Calculated by Quant-It assay until 10^{-2} dilution, lower dilutions extrapolated.
n.a. = no amplification after 45 cycles. n.d. = not done

Limit of detection: MAP cell suspensions. To determine the limit of detection for the MAC subspecies MAP, DNA was extracted from serially diluted MAP ATCC 19698 cell suspensions on two separate occasions and the DNA was used in PCRs using the IS*900*-2 primers. Amplification of the IS*900* target from serially diluted MAP cells yielded similar results to IS*1311* amplification in MAA as shown above. For both MAP extractions, the limit of detection was between 10^3 and 10^4 cells / 100 µL DNA extraction volume.

Triplex assay. Having optimized the components of the multiplex PCR, the EraGen MultiCode RTx system for MAC was tested with the IS*1311*-1, IS*900*-2 and IAC primers combined in a single reaction. This multiplex system allowed detection of MAC DNA and also specified whether MAP DNA was present in a sample. The final triplex PCR assay, the IS*900*-2 forward primer was labeled with the fluorescent marker Cy5, the IS*1311*-1 was FAM-labeled an IAC primers were labeled with HEX. This allowed detection of three amplification targets on three separate color channels / wavelengths. The initial experiment to evaluate this prototype triplex real-time PCR assay used serially diluted MAP ATCC 19698 DNA (100 to 0.001 ng), DNA extracted from serially diluted MAP ATC 19698 cells (approximately 1×10^7 to 3×10^3 cells), and DNA from several non-MAC negative control strains.

In this triplex reaction, both the IS*1311* and IS*900* targets from all MAP samples amplified with a Ct comparable to reactions containing the IS*1311*-1 and IS*900*-2 primers alone. However, there was a problem with the detection of the IAC product on the 640 nm (HEX) channel. There were two melting peaks present, apparently because the Cy5 fluorescent marker bled over into the HEX channel. Approaches to address this problem included checking the light cycling instrument itself and the color compensation files. These attempts reduced but did not eliminate bleed-over into the IAC/HEX channel.

Despite this drawback, the multiplex assay was still functional as the sole purpose of the IAC was to confirm that the PCR was working when there was a negative result for the target of interest. (In other words, when there was a negative signal in Cy5 (IS*900*) and FAM (IS*1311*) channels, amplification of the IAC in HEX channel confirms the PCR was not inhibited.) Even with the bleed-over, it was possible to detect the amplification of the IAC by observing a melting peak at 84°C. Fortunately, there was no bleed-over effect in either the FAM or Cy5 channels so detection (and potentially quantification) of the IS*1311* and IS*900* PCR products was still possible.

PCR B: JTC PCR

Employed alongside PCR A was a non-quantitative multiplex molecular assay previously developed by the Johne's Testing Center. The gel-based assay was designed to distinguish among MAC species isolated through culture from animal tissue and fecal samples. PCR B incorporated five unique sets of primers designed from published insertion sequences of IS*900*, IS*901*, IS*1311*, IS*1245*, and 16S rRNA mycobacterial DNA. The primer sets specific for each insertion sequence, 16S rRNA and the PCR reaction conditions ensured that the target sequences were specific and efficiently amplified. Band size and position on the gel are revealing, with PCR products including a 484 bp product specific for 16S rRNA gene, a 398 bp product from IS*900* specific for MAP, 753 bp and 1290 bp PCR products from IS901 and IS*1245* respectively found in MAA and a 608 bp product from IS*1311* found in all species of the MAC. Prior to this project the specificity of the assay had been confirmed with 76 varied mycobacterial strains including 18 MAP isolates, 9 *M. avium intracellularae* and 10 MAA isolates. Eleven non-mycobacterial species such as *E. coli* and *A. hydrophilia* were tested as negative controls. PCR B results were compared to those provided through high performance liquid chromatography (HPLC). PCR B detected and correctly identified 100% of MAA and MAP. In addition, the assay was not confused by the presence of both MAA and MAP in a single culture and signaled correctly that both species were present. Protocols were optimized for PCR B to be used in this project with spiked water samples to provide a molecular baseline against which to compare the performance of the RT PCR in development (PCR A).

DISCUSSION AND CONCLUSIONS

Real-time PCR primers were optimized to selectively amplify DNA from MAC organisms. The IS*1311*-1 and IS*900*-2 primers included in the final real time PCR (PCR-A) amplified all of the tested MAA and MAP strains. (Unfortunately, since the MAC ITS primers demonstrated sub-optimal performance in the MultiCode RTx system, a primer that would amplify *M. intracellularae* was not produced.)

Both a two target system with IS*1311* and IAC to detect MAC and a three target IS*1311*, IS*900* and IAC to detect MAC and the MAP subspecies are functional. This biplex assay can be used quantitatively, since a linear relationship exists between the amount of target DNA and the data provided by the RT PCR (i.e., the Ct). The limit of detection averaged 10^3 cells / 100 µL DNA extracted. PCR-B, although not quantitative, was found to be useful in detecting and discriminating among MAC species in water samples.

The use of multiple-target PCRs is recommended as an effective approach for samples likely to contain genetically closely related mycobacterial species. Since the pathogenicity across the sub-species in the MAC complex and even between strains of the same sub-species varies significantly, it is important to identify isolated mycobacteria as specifically as possible.

CHAPTER 9
OBJECTIVE SEVEN

OBJECTIVE 7: EVALUATE METHODS OF DETECTING SURFACE-ASSOCIATED AND PLANKTONIC WATER-HABITUATED MAC.

Sub-objectives were:

7a. Compare the growth patterns of planktonic MAA in static vs. moving water chambers.
7b. Evaluate cotton swabbing as a method for recovering MAA from PVC coupon surfaces.
7c. Evaluate the effect of water sample handling and sample environment conditions on MAA enumeration.

METHODS

General Approach. Drinking water system environments were simulated using closed biofilm annular reactors (BARs) (Biosurface Technologies, Model 1320, Bozeman, MT) and miniature reactors (MRs, 40ml I-Chem tube with coupon holder) with autoclaved water from the University of Wisconsin pilot-scale drinking water treatment plant (UW pilot plant) spiked with MAA (ATCC 35712). Each BAR simulated an average one foot per second flow velocity in an equivalent sized 8-inch diameter water main (approximately 30% speed). Each independent MR was rotated on a horizontal axis to induce mixing and simulate variable flow conditions. Matched static water control samples were assessed in independent 150 ml closed polycarbonate control chambers (CC) (Nalgene MF75 Series). BAR and MR coupons, BAR and MR bulk water, and CC bulk water samples were collected for enumeration using the MGIT liquid culture method. To count MAA associated with coupon surfaces and to evaluate coupon surface elution methods, coupon sizes were reduced (to 4.8 mm x 73 mm) to permit their being directly placed into the MGIT incubation tube. This novel direct enumeration method allowed for surface-associated MAA to be counted without cell loss due to harvesting inefficiency.

Source of water. For all experiments, water from the UW pilot plant was collected in a carboy, transported to the lab, transferred to glass bottles and sterilized in an autoclave at 121 °C for 1 hour.

Decontamination. Prior to starting a trial, each BAR was disassembled for maintenance and decontaminated as shown in Appendix 3. MRs were autoclaved at 121 °C for 30 minutes.

Inoculating Protocol. Stock suspensions of late phase exponential growth MAA cells were prepared for inoculation as described in Chapter 1 of this report. After enumeration by flow cytometry, each BAR, MR and CC was inoculated with approximately 10^5 CFUs/ml of MAA.

Trial 1. BAR 1 was held at room temperature (approximately 20 °C) and BAR 2 operated at 37 °C. Both BARs contained 40 polyvinyl chloride (PVC) coupons. Bulk water was sampled weekly for 26 weeks by transferring approximately 4 ml of liquid to a 15 ml polypropylene Falcon tube. After vortexing the Falcon tube for 1 minute, a 0.2 ml sample was transferred to MGIT tubes for enumeration. Periodic sampling of this trial's CCs continued for 80 weeks.

Trial 2. This trial differs from Trial 1 by the type of coupons used, by maintaining each BAR at the same temperature, and by the use of MRs. Each BAR contained 10 PVC, polypropylene (PP), polycarbonate (PC) and polystyrene (PS) coupons. Each MR contained three coupons. Six PVC and four PP, PC and PS MRs were assembled. Coupons were removed from the BAR and MR, and placed directly into the MGIT tubes. For all of Trial 2, the bulk liquid sample of the BAR and MR was inoculated directly to MGIT. Samples were collected for nine weeks. Each BAR, MR and CC for Trial 2 was inoculated with approximately 10^5 MAA CFU/ml and maintained at 37° C. Periodic sampling of the trial's CCs continued for 36 weeks. Additional methods detail is available in Appendix D.

Methods for Each Sub-Objective

7a. Compare the growth patterns of planktonic MAA in static vs. moving water chambers at two temperatures. Samples were collected periodically from the BARs, MRs and the control chambers to compare growth patterns under different conditions.

7b. Evaluate cotton swabbing as a method for recovering MAA from PVC surfaces. The front surface of the coupons was strongly swabbed five times and the swab was placed into a MGIT tube to enumerate harvested MAA cells. The back surface of the coupon was swabbed five times and the swab was placed into a different MGIT tube for enumeration. The swabbed coupon itself was then placed in a MGIT tube to count MAA cells remaining on the coupon after swabbing. Planktonic and surface associated MAA from the BAR that had contained the coupon were counted for comparison. Swabbing of coupons was performed in duplicate. The coupon-associated MAA were counted in duplicate and the planktonic MAA counted in triplicate.

7c. Evaluate the effect of water sample handling and sample environment conditions on MAA enumeration. Organism settling and/or surface attachment was investigated using the CCs by sampling from different chamber locations (depths) prior to and after vortexing. The number of MAA CFU/ml was scaled to the associated volume of the sampling location to obtain a relative distribution of MAA in the control chamber. Also, the effect of processing materials such as different types of pipettes (borosilicate glass or polytetrafluoroethylene) was assessed. For these experiments, bulk water BAR liquid and coupon sampling occurred as described previously. At week 26, to assess conditions that may influence enumeration, additional bulk liquid samples were obtained after 15 minutes of BAR operation at 50% speed and 100% speed (to simulate water distribution system scouring).

RESULTS

Sub-objective 7a: Compare the growth patterns of planktonic MAA in static vs. moving water chambers. Late phase exponential growth MAA inoculated at 10^5 CFU/ml into CCs persisted until the end of the experiment (80 weeks) whether held at 20°C or 37 °C (Figure 9.1). In the CCs, MRs and BARs, MAA was surface-associated after fifteen minutes of contact, and planktonic MAA numbers dropped significantly within three weeks (Figures 9.1 and 9.2). An extrapolation of the CC data predicted that complete die-off would not be reached for many years. At the end of the Trial 1's BAR moving water experiments (26 weeks and after scouring), MAA concentrations were comparable to those in the temperature-matched static water CCs (Figure 9.3).

Sub-objective 7b: Evaluate cotton swabbing as a method for recovering MAA from PVC coupon surfaces. Coupons designed to fit directly into a MGIT tube were effective in directly evaluating how well cotton swabbing removed MAA from coupons for counting (Figure 9.4). Swabbing recovered only 22% of MAA associated with the PVC surface. At least 42% of MAA were lost during processing and 36% were left behind on coupon surfaces.

(Unexpected inactivation of MAA in the BAR containing non-PVC coupons prevented testing these coupons by swabbing.)

Sub-objective 7c: Evaluate the effect of water sample handling and sample environment conditions on MAA enumeration. Results for settlement and surface attachment of MAA in the CCs are shown in Figure 9.5. It appeared that significant numbers of MAA cells were attaching to the surfaces of the BARs, were settling within the BAR, or were being lost through sample handling. After 14 days of static conditions after spiking the water with MAA, a majority of MAA cells were located in the bottom of the CC. In counting MAA cells after shaking the chambers, 68% were found to have been surface-associated prior to agitation. Within 15 minutes after agitation of the control chamber the number of planktonic cells was reduced by 40%. Within 60 minutes the distribution of the cells reverted to the pre-agitation profile: 66% of cells were no longer suspended, comparable to the cell distribution seen in a separate experiment.

No significant influence on cell counts was found between borosilicate glass serological pipettes or polypropylene repeat pipette tips used to harvest CC water samples (Figure 9.5). However, a simple transfer of 4ml of BAR water to a polypropylene Falcon tube prior to inoculation to a MGIT tube recovered roughly 2 log fewer MAA than when the water was inoculated directly into the MGIT tube.

Increased BAR speed by 50% and 100% at week 24 to simulate scouring conditions increased the planktonic MAA cell counts by one order of magnitude as shown in Figure 9.3.

DISCUSSION

Actively growing MAA inoculated into a nutrient poor environment such as water in a distribution system are capable of persisting for an extended period. This project's findings are consistent with previous reports both in experimental systems (Norton 2004; Torvinen, 2007) and in drinking water systems sampled at point of use (Hilborn 2006).

A mass balance analysis approach (i.e., calculating the total number of MAA CFUs by volume) did not explain the observed control chamber data. A kinetic relationship, as opposed to an exponential decay function, best described what occurred with planktonic MAA in this environment. The shape of this kinetic relationship is a rectangular parabola, similar to that found with Monod kinetics. This kinetic relationship suggests that the number of replicating MAA was in some way dependent upon the number of MAA inactivated, raising the possibility that replicating MAA could be using dead or dormant MAA as a substrate source. (If inactivation only were occurring, an exponential decay function would have been seen).

The waxy cell wall of these organisms results in MAA's tendency to repel from water and associate with hydrophobic surfaces. This tendency contributes to biofilm formation. To better investigate MAA hydrophobicity and the extent of its association with surfaces, an efficient method of removing MAA is needed. Cotton swabbing was ineffective and enumeration relying upon this method likely will underestimate MAA cell numbers. While this observation is specific to the cotton swab method, based on previous biofilm elution experiments it appears applicable to scraping methods as well. (Scraping methods were not evaluated in this study since placement of a metal blade in MGIT media would interfere with oxygen detection results through oxidation of the blade.)

An analysis of surface association data suggested that after 14 days most, but not all, MAA cells were located at the bottom of the CC or surface associated prior to agitation. The number of MAA recoverable varied by chamber depth prior to agitation, and significantly increased afterwards. It is possible that this depth-related variation is due to smaller clumps of MAA staying in suspension, with settled clumps of various sizes at the bottom of the chamber. Another observation was that BAR surface attachment and/or settling patterns were affected by conditions mimicking scouring. A one log greater MAA concentration was seen in samples collected after scouring than before. These findings together suggest a favorable sampling time for MAA surveillance in water distribution systems is during routine flushing of mains.

Transient contact with sampling materials is unlikely to influence MAA enumeration (i.e., water samples drawn up and immediately expelled from a pipette). However, extended contact of MAA with polypropylene (a hydrophobic material much like PVC commonly used for collection and processing tubes) should be expected to result in a loss of planktonic MAA. The organisms quickly associate with the material's surface much as MAA associated to PVC coupons in the BAR or to PVC, PP, PC and PS in the MRs, and likely remain associated until scoured off or physically removed.

CONCLUSIONS

The persistence patterns shown here have been previously observed from a qualitative perspective using molecular methods with EM with starved organisms (Archuleta et al., 2005), and with *M. tuberculosis* in response to oxygen depletion (Wayne and Hayes, 1996). A kinetic relationship based on time after inoculation provides a good overall fit to the MAA recovery data.

Since comparable growth patterns were seen in both CCs, it appears that at the incubation temperatures studied (20 ºC vs. 37ºC) temperature had limited effect on the growth patterns of water-habituated MAA.

The BAR results indicate that once MAA were associated with a hydrophobic coupon surface such as PVC, PP, PC and PS, they remained associated until they are scoured off. This pattern was not seen with water-habituated planktonic MAA in CCs that were not allowed to remain associated with a surface. Instead, MAA in the CCs show a steep initial decline in CFUs/ml with time. Surface association may be a preferred mode of habituation for MAA in water distribution systems.

1. MAA spiked into autoclaved water died off neither in a linear nor exponential pattern but in a persistence pattern best described as a kinetic function.
2. Water-habituated MAA associated to surfaces suggesting that under normal flow conditions in a drinking water distribution system the majority are not planktonic.
3. Hydrophobic properties of MAA, assumed to be consistent across MAC species, could interfere with enumeration attempts if samples are not properly obtained or handled. Further study is needed to develop an optimal sample collection method which recognize these properties to address handling and holding protocols, bulk water and biofilm collection and sample concentration methods. Further study using single sample mini-chambers with coupons would permit unperturbed sampling, limit chances for contamination and minimize organism loss through handling.
4. Growth patterns of planktonic MAA at 20ºC to 37ºC did not differ.
5. Growth patterns of planktonic MAA in static vs. moving water may not differ. Further study is needed after enhancement of sample collection protocols (use of single sample miniature chambers) to determine if differences exist.

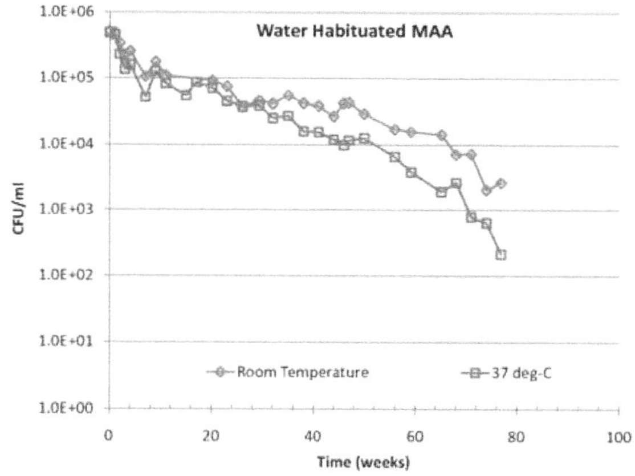

Figure 9.1. Water-Habituated MAA Enumeration Results for 80 Weeks at Two Temperatures in Static Water Control Chambers

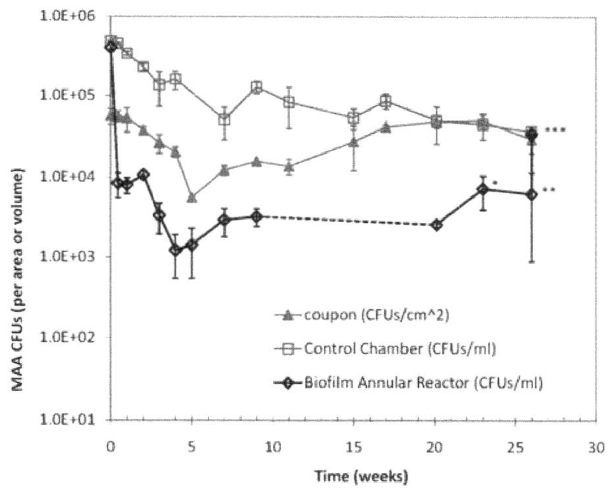

Figure 9.2 Comparison of PVC Coupon-Associated MAA with Planktonic MAA in BAR2 and Matched CC. *Direct inoculation to MGIT tube (vs. passage through one collection vessel) Week 26: **sample before BAR scouring. ***sample after BAR scouring.

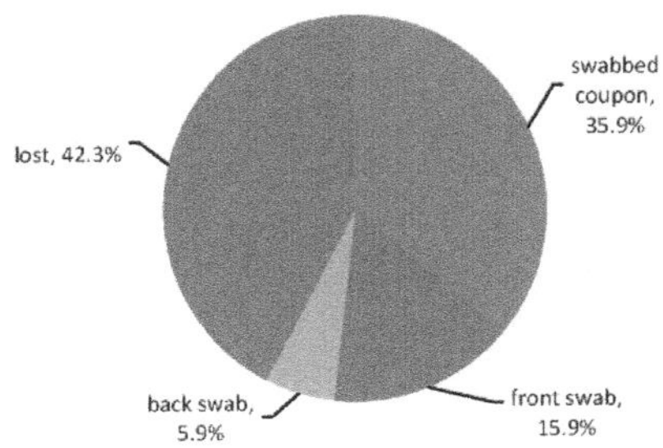

Figure 9.3. MAA Recovery from PVC Coupons by Cotton Swabbing.

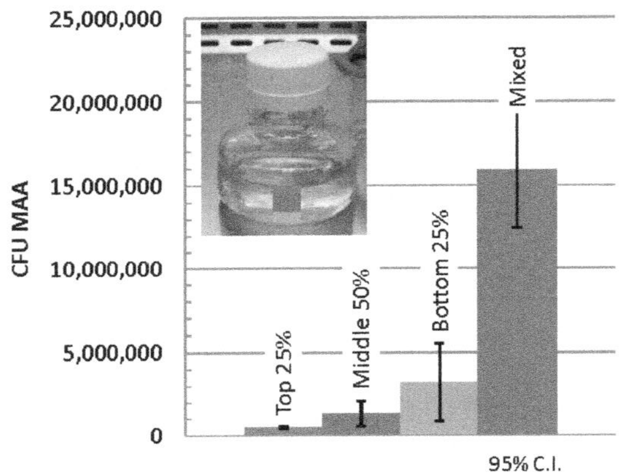

Figure 9.4. Partitioning in Control Chamber After 14 Days. CFU = colony forming unit; C.I. = confidence interval.

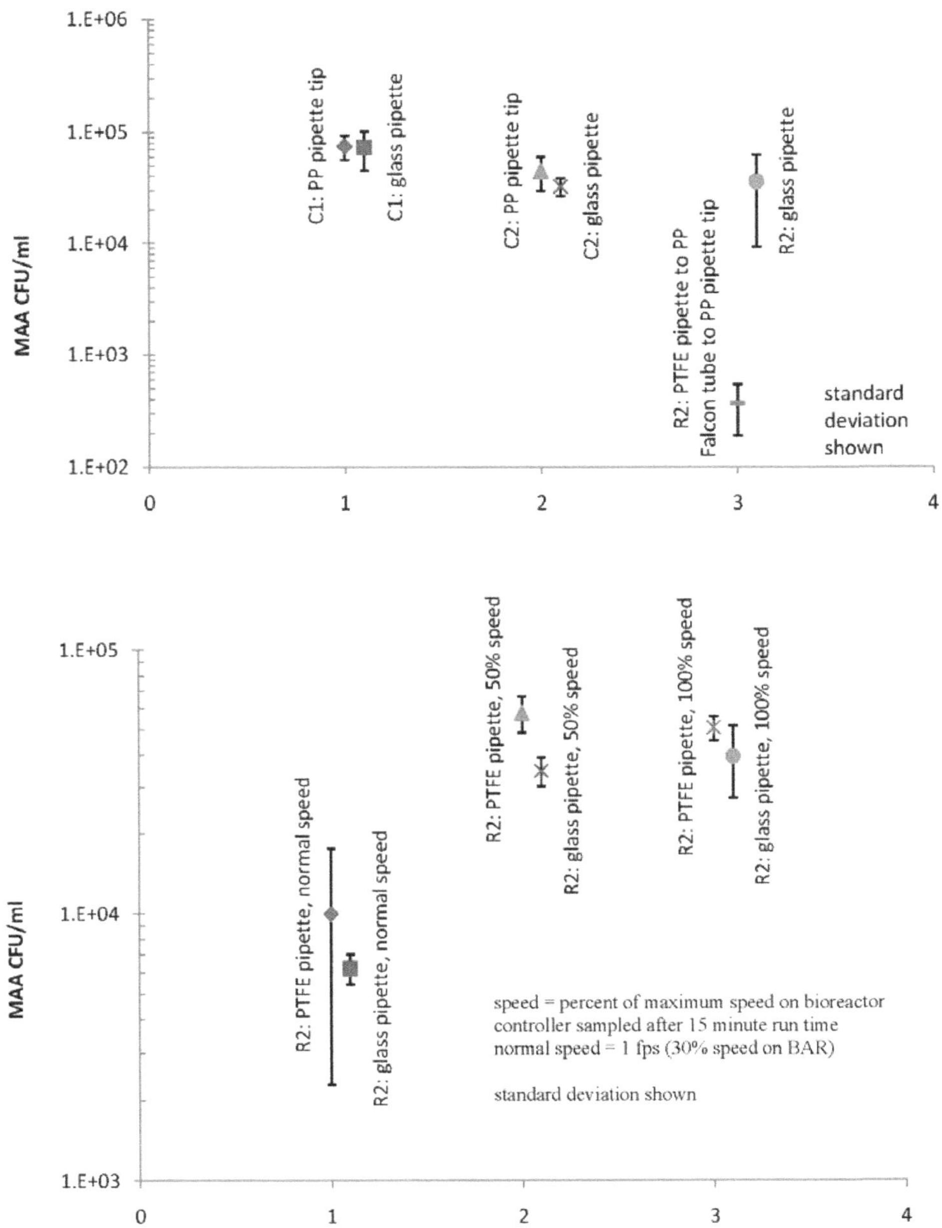

Figure 9.5. Biofilm Annular Reactor (BAR) Surface Attachment Analysis: Sampling Methods and Materials. R = BAR, C = Control Chamber (CC), 1 or 2 indicates CC or BAR number.

CHAPTER 10
FINDINGS AND RECOMMENDTAIONS

The following summarizes the conclusions obtained from the seven objectives of this study:

1. The method providing the greatest sensitivity in screening water for residual MAC species after treatment couples liquid culture-based methods with PCR. Combining the two approaches compounds their advantages while diluting their disadvantages. The culture-based method using the MGIT 960 system amplifies the number of organisms for subsequent DNA extraction, thus maximizing sensitivity while simultaneously limiting the effect of PCR inhibitors. This method was shown capable of isolating MAC out of a finished water sample from the Pinellas Utility distribution system.

2. The MGIT culture method also provides a way to enumerate the MAC cells in a sample. This project developed the first algorithm to convert MGIT 960 instrument Time-To-Detection (TTD) values to Colony Forming Units (CFU) for any strain of MAP and has also established MAA strain-specific algorithms to convert TTD to CFU. The method has proven valuable as a research tool for rapid and cost-effective MAC enumeration. The utility of the method on finished water remains to be determined since the diversity of mycobacteria in some sample types could complicate MAC quantification by MGIT methods.

3. Use of free chlorine at standard concentrations to minimize contaminating non-mycobacterial organisms is unlikely to interfere with isolation of MAC species. Inclusion of the commonly used decontaminant CPC in water testing protocols may augment hydrophobic partitioning of mycobacteria and interfere with accurate MAC quantification.

4. Decontamination of water samples using CPC has little or no effect on MAC viability but substantially reduces the number of viable non-mycobacteria in samples enhancing their recovery by culture/PCR methods.

5. Filtration, followed by DNA extraction from the filter, is superior to centrifugation as a method of concentration and identification of MAC species from finished water samples but significant losses of MAC from samples occur due to hydrophobic interactions with filtration equipment, pipettes and/or other plasticware confounding assays.

6. Commercially available antibodies to MAC were not effective for immunoconcentration techniques but novel antibodies produced by the project were highly effective at selective concentration of MAA from water samples. However, technical problems prevented use of these antibodies with immunomagnetic beads. Alternative formats providing greater surface area for attachment, for instance polystyrene trays coated by the antibodies such as those produced during this project, are promising.

7. The triplex PCR method, which can be performed even in the presence of contaminating organisms, quickly identifies the presence of mycobacterial species in general as well as

specific MAC subspecies based on novel genetic targets and also provides a degree of quantification using real-time PCR technology.

8. Direct quantification of MAC in biofilms by placement of coupons directly into culture media was superior to quantification in biofilms harvested by scraping or cotton swabs.

9. The hydrophobic properties of MAC have a significant impact on quantification methods, i.e., substantial numbers attach to plastic surfaces and appear to "disappear" from assays (see conclusion #4 above). The organisms' location and behavior in water systems, e.g., rapid attachment to solid surfaces, are dependent on the hydrophobicity of those surfaces. This fundamental finding needs further study as it significantly affects future MAC study designs.

RECOMMENDATIONS FOR UTILITIES

For utilities conducting surveillance for MAC organisms:

1. Samples should be obtained during flushing of water mains to collect water representative of removed sediment and biofilm.
2. A secondary method of sampling and analysis is to suspend PVC coupons in the flow stream during flushing to select for hydrophobic organisms and minimize decontamination requirements. This method would be optimal for MGIT liquid culture detection, with mycobacterial isolates identified present by molecular methods.
3. To minimize loss of organisms from any fluid sample after collection, hydrophilic materials such as Teflon coated or glass chambers be utilized for sample collection. Efforts should be made to minimize handling and transfer of sample among vessels.
4. Analysis of a collected suspended bulk water sample should occur as soon as possible after efforts to resuspend them to minimize loss of sample associated with settling or partitioning.
5. Given the typically low numbers of MAC in treated water, detection attempts should rely on assays with the greatest sensitivity. Use of both culture followed by PCR and direct PCR of samples will yield the most accurate assessment of the number of MAC organisms contaminating drinking water.

Future research. Methods that minimize loss of these hydrophobic organisms through sampling materials and handling in order to ensure accurate finished water surveillance should be developed.

APPENDIX A

Chapter Four: Filtration vs. Centrifugation Methods Detail

Verification of water spiking method. Six 1 L polycarbonate bottles of autoclaved Madison tap water were inoculated with approximately 1 x 10^6 MAA cells. The inoculated water was vigorously shaken by hand for 1 minute. Three 200µl aliquots were collected from each bottle for MAA counting by MGIT to establish the actual initial MAA cell count in the spiked water samples.

Enumeration. Counting MAC cells using MGIT TTD was based on the method established by Shin (Shin et al., 2007). Briefly, a serial dilution of late exponential phase growth MAA was concurrently inoculated to MGIT and plated to solid medium. Based on plate counts, and an assumed one order of magnitude difference between each dilution, an equation to fit MGIT TTD with inoculate size was developed. For MAA strains, a log linear standard curve relationship was found as described:

$$X_{cfu} = 10^{-0.5425*ttd+7.5995}$$

Where

X_{cfu} = number of colony forming units in the MGIT tube
ttd = time to detection from MGIT instrument

Centrifugation. Three 1 L polycarbonate bottles of autoclaved Madison tap water were inoculated with approximately 1 x 10^6 MAA cells. The inoculated water was vigorously shaken by hand for 1 minute. The spiked water was transferred to five, 250 ml plastic bottles and centrifuged at 4000 x g for 25 min at 6°C. The resulting pellet from each 250 mL sample was harvested and combined in a single sixth plastic centrifuge plastic bottle along with 10mL of PBS used to rinse each of the original five bottles. The sixth bottle containing the composite harvest from the first 5 bottles was again centrifuged at 4000 x g for 25 min at 6°C. The final pellet was resuspended in 1 ml of PBS buffer and transferred to a 10 mL plastic Falcon tube. Three, 100µl aliquots were removed and inoculated into MGIT ParaTB medium to establish the viable MAA count.

Trial 1a: test the effect of *centrifugation force and duration* on MAA recovery. Two additional liters of autoclaved Madison tap water were inoculated with 2 x 10^6 MAA ATCC 35712 cells. This spiked tap water was then divided into 10, 200 ml aliquots and placed in 250 ml sterile plastic centrifuge bottles. The bottles were centrifuged at one of three speeds (2830 x g, 4420 x g, or 12800 x g) for either 30 or 60 min. Each centrifuge condition was done in duplicate. After centrifugation, the supernatant was decanted and the pellet was resuspended by vortexing in 5 ml of PBS buffer. Three, 100µl aliquots of each resuspended pellet were inoculated to MGIT tubes of ParaTB medium to determine the number of MAA cells recovered.

Trial 1b: test the effect of using *fewer sample containers for centrifugation*. High speed centrifugation (12,800 x g for 25 min) was employed using a method of water processing that minimized the number of centrifuge bottles. Three, 1 L samples of autoclaved Madison, WI tap water in 1 L glass bottles were inoculated with approximately 2 x 10^7 MAA ATCC 35712 cells. An additional 1 L Madison, WI tap water sample was left uninoculated as a negative control. The inoculated and uninoculated water samples were transferred to 500 ml centrifuge bottles (Nalgene polycarbonate, # 05-564-2) in 400 ml aliquots. Samples were centrifuged at 12,800 x g for 25min. The supernatant was discarded and an additional 400 ml of sample was added to each centrifuge bottle and centrifuged (12,800 x g for 25 min). This was repeated for a total of three times until the full original 1 L water sample had been processed. After the last decanting of water, each original glass sample bottle was rinsed twice with 100 mL phosphate buffered saline (PBS) and the rinse was added to the centrifuge bottle and centrifuged one last time. The final cell pellet (which was not visible) was resuspended in 5 ml PBS buffer. Three, 300µl aliquots of the resuspended pellet were transferred to MGIT tubes of ParaTB medium to determine the number of MAA cells recovered. As an additional measure of MAA recovery total DNA was quantified. Briefly, 300µl aliquots were transferred to ribolyser tubes for DNA extraction using the method outlined in Appendix 1. The extracted DNA was quantified using the Quant-iT™ High Sensitivity DNA Assay Kit (Invitrogen, Carlsbad, CA).

Filtration. Threel L polycarbonate bottles of autoclaved Madison tap water were inoculated with approximately 1 x 10^6 MAA cells. The inoculated water was vigorously shaken by hand for 1 minute. An additional 1 L water sample was left uninoculated as a negative control. Polycarbonate track etched 0.4 µm filters (47 mm, Millipore Isopore type HTTP) were placed on autoclaved, glass filter holders. Triplicate 1 L samples of the MAA spiked tap water were filtered through both the 0.4 µm filters after they were rinsed with 100 ml sterile distilled water with 0.01% Tween 80. After the 1 L sample was filtered, the sample bottle and filter was rinsed with 100 ml sterile, distilled water. The filters were placed directly into MGIT ParaTB Medium tubes to determine the number of MAA cells captured on each filter.

Trial 2a focused on *recovery of MAA from PCTE filters by sonication.* Four, 1 L samples of MAA spiked tap water were filtered through 0.4 µm pore-size PCTE filters. The filters were then placed in glass jars containing 5 ml PBS with 0.05% Tween 80. The filters were sonicated in a bath sonicator (Branson, 60 Hz, 117V) for 5 or 10 minutes (duplicate samples for each time). After sonication, the filters were removed from the buffer, rinsed briefly in 20 mL sterile water and the number of MAA counted by MGIT methods (100 µL inoculum). Previous experiments showed that this manner of sonication did not affect the viability of MAA (data not shown).

For Trial 2b, *acetone treatment of filters,* dissolvable cellulose acetate filters that have been used to recover *Cryptosporidium* oocysts from water (Aldom and Chagla, 1995) were employed. After dissolving the membranes with acetone the oocysts are sequentially washed in ethanol and distilled water (Aldom and Chagla, 1995). To determine if this procedure might be used for MAA recovery, the sensitivity of MAA to acetone and ethanol was tested. Approximately 1 x 10^6 MAA cells were exposed to acetone for two minutes, centrifuged and washed with 95% ethanol, 70% ethanol and twice with water with centrifugation at 2,500 x g for 15 min between each wash. As controls, additional MAA cells were washed only in ethanol and water as described above without acetone treatment. At the end of the last wash step, cells were resuspended in 500 µl PBS buffer and triplicate 100µl aliquots were placed in MGIT tubes to determine the number of viable cells remaining after the procedure.

To test the effect of PCTE *filter pore size* in MAA recovery for Trial 2c, ten 1.0 L glass bottles of autoclaved Madison, WI tap water were each spiked with approximately 3 x 10^7 MAA ATCC 35712 cells. The inoculated water was vigorously shaken, by hand, for 1 minute. Two, 100µl aliquots from each bottle of water were inoculate to tubes of MGIT ParaTB medium to establish initial MAA cell counts. For filtration, 0.4 µm (47 mm, Millipore Isopore type HTTP) or 0.2 µm (47 mm, Millipore Isopore type GTTP) were placed on autoclaved, glass filter holders. Each bottle of spiked water was filtered through either a 0.4 µm or 0.2 µm pore size filter. The filters were initially rinsed with 100 mL sterile distilled water with 0.01% Tween 80. After the filtration, the sample bottle and filter were rinsed with 100 mL sterile, distilled water. The filters were placed directly into MGIT tubes to determine the number of MAA cells captured on the filters. Two, 200µl aliquots were taken from the filter effluent to determine the number of organisms that may have passed through the filter.

For Trial 2d, four *different filter types* were compared: polycarbonate track etched filter (PCTE); mixed cellulose acetate cellulose nitrate filters (cellulose, Fisher Scientific 09-719-2E); nylon (Whatman 7404-004); and Durapore™, a hydrophobic polyvinylidene fluoride membrane (Millipore, HVLP 04700). All filters tested were 0.4 µm pore size.

MAA cells were inoculated to four glass bottles each containing sterile 1 L Madison, WI tap water. After thorough mixing, the MAA cell concentration in the spiked tap water was determined by the MGIT counting method (300µl of water was inoculated into tubes of MGIT ParaTB medium). Each of the four 1 L spiked tap water samples was filtered using one of the four filter types. Each sample bottle was rinsed twice with PBS and the rinse buffer was passed through the same filter. The filters were transferred directly to tubes of MGIT ParaTB medium to calculate the total number of cells collected on each filters. The Durapore™ membranes were pre-wetted with 1 ml of 70% EtOH and rinsed with 10 ml dH_2O prior to sample filtration. A vacuum regulator was used to hold the vacuum pressure to between 10 - 15 mmHg. This vacuum pressure also was used for all subsequent filtration experiments.

To assess relative DNA recovery (vs. whole cell recovery described above) from different filter types, three additional 1 L samples of MAA-inoculated tap water were filtered through PCTE and nylon filters as was 1 L of uninoculated water. These filters were transferred to ribolyser tubes for DNA extraction. The resulting DNA extracts were used in real-time IS*1311* PCR assays using standard conditions and a 62°C annealing temperature. The DNA in the samples was quantified using the Quant-iT™ High Sensitivity DNA Assay Kit (Invitrogen).

In addition, three 200 mL samples of PBS in glass bottles were inoculated with MAA ATCC 35712 cells. The PBS was tested for viable MAA, total DNA and IS*1311* both before filtration (MAA cells in suspension) and after filtration (MAA cells trapped on the 0.4 µm PCTE filters). Real-time PCR results for IS*1311* were expressed semi-quantitatively as average Ct values.

A plausible explanation for loss of MAA cells during filtration concentration experiments is passage of the MAA cells through the 0.4 µm filter, i.e. filter break-through. In several MAA filter concentration experiments, 300 µl of the filtrate (water that passed through the filter) was transferred to tubes.

Trial 2e investigated the *adherence of MAA cells to glass.* One liter of sterile Madison, WI tap water was inoculated with $2x10^5$ MAA ATCC 35312. Two 200 mL aliquots were filtered using untreated glass filter bells. After filtration, the filters were transferred MGIT culture tubes to determine the number of cells collected on the filter. The base of the glass filter bell (~31 cm^2 area out of ~207 cm^2 total surface area) was then swabbed with a sterile cotton swab and the swab then transferred directly to a tube of MGIT ParaTB medium to assess the number of MAA cells adherent to the glass filter bell.

Trial 2f tested a blocking agent to minimize MAA cell loss since prior filtration trials had significant, i.e. 80% or more, loss of MAA cells. The working hypothesis was that these cells were attaching to the wall of plastic or glass vessels used to carry out the experiments. Sigmacote™ (Sigma-Aldrich, St. Louis, MO), a silicone solution in heptane that forms a covalent microscopically thin film on glass, was investigated as an approach to block MAA cells from adhering to surfaces in contact with the MAA-containing water.

Sigmacote™ (Sigma- Aldrich, St. Louis, MO) was applied to glass filter bells as per manufacturer's instructions. One liter of sterile Madison, WI tap water was inoculated with $2x10^5$ MAA ATCC 35312. Two 200 mL aliquots were filtered using glass filter bells coated with Sigmacote and two 200 mL aliquots were filtered using untreated filter bells. After filtration, the filters were transferred directly to tubes of MGIT ParaTB medium to determine the number of cells collected on the filter.

APPENDIX B

Chapter Six: Evaluation of anti-MAC antibodies detail

Cell suspensions were enumerated by flow cytometry and either 5 x 10^5 or 1 x 10^6 cells were aliquoted into 1.7 mL microcentrifuge tubes. The bacteria were pelleted at 14,000 x g and resuspended in primary antibody that had been diluted in PBS with 0.1- 1.0% bovine serum albumin (BSA). Cells were incubated 30-60 minutes at room temperature or 37ºC, washed 2-3 times with reagent grade water or phosphate buffered saline to remove unbound primary antibody, and stained at room temperature or 37ºC for 30 – 60 minutes with an FITC conjugated secondary antibody (anti-IgG-FITC, KPL, Gaithersburg, MD or anti-IgY-FITC, Promega, Madison, WI). The MAC cells were washed a final time and resuspended in 0.5 mL PBS.

APPENDIX C

Chapter Eight: PCR Methods Detail

Optimization of MAC ITS primers. An initial PCR target was the internal transcribed spacer (ITS) region of MAC. The primers designed to amplify the ITS region of all MAC strains showed good MAC strain specificity for a limited number of test strains. However, these primers (MAC F1 and R1), used in the MultiCode RTx real-time PCRs, resulted in a relatively high crossing threshold (Ct), i.e. >30, for 50 ng of positive control DNA, suggesting low analytical sensitivity. Attempts were made to optimize the MAC ITS primers to lower the Ct and increase analytical sensitivity.

Optimizing annealing temperature, primer concentration, and extension times. An annealing temperature 61˚C was determined to be optimal for the MAC ITS primers, allowing DNA amplification from most MAC strains but not from other bacteria. Lower annealing temperatures resulted in lower Ct values for MAC strains, however this also allowed amplification from non-target organisms.

Increasing MAC forward and reverse primer concentrations up to 0.6 μM (from the standard 0.2 μM) did significantly decrease the Ct for positive control DNA. Increasing primer concentration further, up to 0.8 μM, resulted in amplification from non-target DNA. All further PCRs used 0.6 μM of the MAC ITS primers.

In another experiment, the Mg^{++} concentration was increased in the PCR master mix in 0.5 μM increments from 2.0 μM (the standard concentration in the EraGen master mix) to 4.0 μM. Increasing the Mg^{++} decreased real-time PCR Ct by approximately 4 cycles for the positive control DNA, however negative control DNA also amplified when the Mg^{++} concentration was 2.5 μM Mg^{++} (data not shown). Therefore, the standard 2.0 μM Mg^{++} concentration was maintained in subsequent trials.

The final PCR optimization element addressed increasing primer extension time from the standard 20 seconds to 40 and 60 seconds. Increased extension time to 40 sec decreased the Ct for positive controls by 1-2 cycles, however it also resulted in amplification of negative control DNA that did not amplify with a 20 sec extension time. With a 60 second extension time, the Ct for the positive control DNA actually increased. As a result of these experiments, the extension time was kept at the standard 20 seconds in subsequent trials.

Testing optimized protocols with MAC repository isolates. With the optimized PCR conditions (61˚C annealing temperature, 0.6 μM primer concentration, 2.0 μM Mg^{++} in the master mix), the specificity of the MAC F1 and R1 primers to DNA isolated from multiple bacterial species was evaluated.

These primers amplified DNA from most of the MAC species tested however three of the MAA strains did not amplify and two non-MAC strains, *M. asiaticum* ATCC 25274 and *M. scrofulaceum* ATCC 19981 did amplify. The Ct for most of the DNA samples were relatively high, suggesting these primers would not be very sensitive for amplifying low concentrations of DNA from these MAC organisms. Given these sub-optimal results, it was concluded that MAC ITS F1 and R1 would not be useful.

Redesign of MAC primers. The MAC ITS primers were redesigned to provide better specificity and sensitivity. The redesigned MAC forward primer (F2) was shifted 4 basepair from the original F1 primer. Two additional reverse primers were also designed: one primer extending one 5' basepair past the original primer (R long, or RL) and one one 5' basepair shorter than the original (R short, or RS). These three primers were synthesized and used in different combinations with the original F1 and R1 primers. The goal was to determine if any of the primer combinations allowed amplification from all the MAA strains with lower Ct values.

Initial evaluation of the F2 primer suggested there was no substantial difference between F1 and F2 primers. The F2 primer in combination with the three reverse primers, (R1, RL, RS) did not result in amplification of the MAA ATCC 35712 or MAP JTC 1319 strains, nor was the Ct decreased with an annealing temperature of 60°C. If the annealing temperature was lowered to 58°C, there was amplification from *M. terrae* and *M. malmoese* negative control DNA, confirming previous results that the annealing temperature needs to be higher to achieve reaction specificity. The RS primer in combination with F1 did not perform well. At the 61°C annealing temperature, MAA EPA-1 and MAP ATCC19698 positive control DNA did not amplify. The RL primer performed as well as the original R1 primer.

This primer was chosen for testing, in combination with the F2 primer, with a larger number of test DNA samples. With an annealing temperature of 61°C, the F2 + RL combination had the same analytical specificity as the original primers set; the MAA 35712, MAP JTC 1319 and MAP JTC 1357 strains did not amplify. In general, the Ct for the positive control DNA samples decreased with the F2 + RL combination, but only by 1-5 cycles.

Limit of detection for MAC ITS primers. A high crossing threshold for 50ng DNA tested was believed to indicate a low assay analytical sensitivity for MAC DNA detection. To test this assumption, PCRs using a serial dilution of MAP ATCC19698 template DNA (from 100 to 0.001ng) were done using the F2 and RL MAC ITS primers (0.6 μM primers, 61°C annealing temperature). From this experiment, the limit of detection for the MAC ITS primers for MAP ATCC 19698 was 0.1 ng DNA or approximately 10^5 cells (Table A.1). This limit of detection was much higher than for two other primers used in this study (IS*1311*-1 and IS*900*-2), which had a limit of detection of <0.001 ng DNA (approximately 10^3 MAP cells).

Evaluation of ribolyser DNA extraction method. A ribolyser (bead-beating) method to extract DNA from MAC cells yielded significantly higher amounts of DNA than other methods evaluated.

An experiment was prepared to evaluate the efficiency with which the ribolyser method lysed MAA and MAP cells. Six 100 μL aliquots of both MAA and MAP cell suspensions were transferred to six ribolyser tubes. For each of the MAC strains, two cell suspension concentrations were used (approximately 10^4 and 10^2 cells per 100 μL sample). For no-treatment control samples, 500 μL of TE buffer was added to triplicate ribolyser tubes for each of the two strains and two cells concentrations. These tubes were held at room temperature for 2.5 hours. For the other set of triplicate ribolyser tubes for each of the two strains and two cell concentrations, 500 μL of DNA lysis buffer was added to each tube. These samples were incubated at 37°C for 2 hours, ribolysed, and held on ice for 15 min. The entire contents of the ribolyser tubes, both no-treatment controls and ribolysed samples, were transferred to MGIT tubes to measure the number of viable cells present in the samples.

For both MAA and MAP cells, greater than 95% of the cells were killed (presumably lysed) with the ribolyser procedure. For MAA, killing efficiency of the ribolyser method was not significantly different ($p<0.01$) at the two cell concentrations used. For the 1 x 10^3 MAP samples, the ribolysed samples remained MGIT culture-negative (up to day 36), indicating no viable cells were present in the sample. These results indicated that the ribolyser DNA extraction method effectively killed, and presumably lysed, the majority of MAC cells.

This experiment was repeated using serially diluted cell suspensions from 10^1 to 10^5 cells / 100 µl sample for both MAA and MAP. In this experiment, there was no apparent difference in ribolyser-induced killing (lysis) efficiencies between the different cell concentrations of either of the MAC strains tested (data not shown). These results indicate lysis efficiencies of the ribolyser method were not likely affected by cell concentration.

Table A.1. Comparison of PCR Limit of Detection (PCR Cycles) for MAC ITS, IS*1311*-1 and IS*900*-2 Primers Using MAP DNA

Sample (ng DNA)	IS1311	IS900	MAC ITS
100	15.8	21.6	28.5
10	18.8	15.2	33
1	22.4	18.4	36.5
0.1	27.8	23.0	42.7
0.01	31.3	26.5	n.a.
0.001	34.4	26.5	n.a.

MAP MGIT counts	IS1311	IS900	MAC ITS
$1.1 \pm 0.2 \times 10^6$	22.3±0.7	18.0±0.7	36.6±1.8
$2.4 \pm 0.2 \times 10^5$	26.3±0.3	21.7±0.6	43.3±0.4
$3.1 \pm 0.2 \times 10^4$	30.6±0.4	25.7±0.6	n.a.
$1.7 \pm 0.2 \times 10^3$	38.6±1.5	32.5±1.1	n.a.
$3.9 \pm 0.2 \times 10^1$	n.a.	n.a.	n.a.

[1] No amplification within 45 cycles

[2] Number of cells in sample for initial extraction

A significant additional problem was "primer dimer" formation that interfered with assay interpretation. Due to these assay performance issues, further work using these primers was not pursued. At the conclusion of these experiments, the MAC ITS region was abandoned as a target for the real-time PCR and work focused on insertion sequence targets as discussed in Chapter 8.

APPENDIX D

Chapter Nine: Methods Detail

Decontamination of each BAR consisted of:

1. Thorough BAR cleaning and autoclaving for 60 minutes at 121 °C.
2. Soaking coupons and coupon holders for 24 hours in a 500 ml/L sodium hypochlorite solution adjusted to pH 6.7 (HCl addition).
3. Aseptically inserting holders and coupons into each BAR. To ensure BAR did not become contaminated while inserting coupons, each BAR was filled with sterilized UW pilot plant water containing 500 ppm sodium hypochlorite adjusted to pH 6.7 and operated at approximately 30% of full rotational speed for 24 hours.
4. Each BAR was aseptically drained and rinsed twice before filling with sterilized UW pilot plant water. The first rinse was performed with autoclaved de-ionized water. The second rinse used sterilized UW pilot plant water.
5. The BAR filled with sterilized UW pilot plant water was operated at approximately 30% of full rotational speed for 24 hours and then sampled for contamination by plating 1 ml of BAR liquid to a blood agar plate incubated at 37°C for 24 hours.
6. If any BAR in a trial proved to be contaminated, the decontamination protocol was repeated.

Inoculation and Sampling Protocol. Approximately 10^5 CFUs per ml of MAA were inoculated to each BAR and CC. After each inoculation was complete, 1 mL of prepared cells and 1 mL of liquid from each BAR and CC was sampled individually to blood agar plates and incubated at 37°C for 24 hours to check for contamination.

Sampling was performed in a biosafety cabinet to maintain sterility of each pure culture environment. Prior to BAR sampling, the exterior of the reactor was decontaminated by spraying with 70% ethanol and allowed to evaporate. After evaporation, the reactor surface was coated with a 500 mg/L sodium hypochlorite solution, pH adjusted to 6.7, and allowed to soak for approximately one minute. Excess NaOCl present around the BAR access port was wiped with a sterile cotton swab before opening to limit entrance of decontaminant into the reactor.

After bulk liquid samples were obtained from each BAR, coupon samples were collected. Coupons were removed from the coupon holders using sterile instruments. Each coupon removed was placed directly into a prepared tube of MGIT ParaTB medium for enumeration. Coupon sampling occurred in duplicate or triplicate for each sampling time as indicated in the results section. As a control, a triplicate trial of a sterile coupon placed in a prepared tube of MGIT ParaTB medium was performed to determine if coupon materials influenced MGIT results.

Each BAR and corresponding CC were sampled at the same time. Because CCs were stationary between sampling times, each CC was vortexed for 1 to 2 minutes to resuspend cells prior to sampling, with samples obtained within 1 minute after completion of chamber vortexing. CC samples were transferred to MGITs in triplicate at each sampling time. To minimize the time between turning off the reactor and obtaining the sample, sampling occurred within 5 minutes of the reactor shutdown.

Coupon swabbing protocol. When duplicate BAR PVC coupons were sampled, two additional PVC coupons were removed and swabbed in the following manner:

1. The front of the removed coupon was swabbed five times and the swab was placed into a MGIT tube to enumerate cells. The back of the coupon was swabbed five times and the swab was placed into a different MGIT tube.
2. The swabbed coupon itself was placed in MGIT to assess cells remaining on the coupon surfaces.
3. Normal BAR sampling with direct placement of coupons to MGIT was used to assess the total number of organisms of untouched coupon surfaces.

Partitioning experiment methods. To investigate the possibility that MAA was settling in the BAR, an experiment was performed to determine the extent of settlement within a control chamber that occurred between sampling times. In this experiment, samples were taken in triplicate from the control chamber as follows: upper 25% of the volume, midpoint of the volume, and bottom 25% of the volume prior to vortexing, and midpoint of chamber after vortexing to resuspend all cells into solution. Enumeration results (MAA CFU/ml) were scaled to the associated volume of the sampling location to obtain a relative distribution of MAA in the control chamber. To further investigate the settlement of MAA in the BAR, an experiment was performed to determine the rate of settlement within a control chamber that occurred after agitation. In this experiment, samples were taken in triplicate from the control chamber after agitation and 15, 30 and 60 minutes afterwards. The initial sampling site after agitation was the midpoint of the chamber. For lapsed time samples, the two sampling sites were midpoint of upper 50% of the volume and midpoint of bottom 50% of the volume. As part of the partitioning experiment with BAR 2 described above, the effect of pipette material was further assessed. For these experiments, bulk water liquid and coupon sampling occurred as normal, with additional bulk liquid samples obtained after 15 minutes of operation at 50% BAR speed and 100% BAR speed, each of which simulated scouring conditions within a water distribution system. Bulk water samples were obtained with borosilicate glass or polytetrafluoroethylene (PTFE) serological pipettes.

REFERENCES

Abubakar, I., D. Myhill, S.H. Aliyu, P.R. Hunter. 2007. Detection of *Mycobacterium avium* subspecies *paratuberculosis* from patients with Crohn's disease using nucleic acid-based techniques: a systematic review and meta-analysis. *Inflamm. Bowel Dis.* 14:401-410.

Aldom, J.E. and A.H. Chagla. 1995. Recovery of *Cryptosporidium* oocysts from water by a membrane filter dissolution method. *Letters Appl. Microbiol.* 20:186-187.

Archuleta, R.J., P. Hoppes, Y. and T.P. Primm. 2005. *Mycobacterium avium* enters a state of metabolic dormancy in response to starvation. *Tuberculosis* 85:147-158.

Aronson, T., A. Holtzman, N. Glover, M. Boian. S. Froman, O.G. Berlin, H. Hill, and G. Stelma, Jr. 1999. Comparison of large restriction fragments of *Mycobacterium avium* isolates recovered from AIDS and non-AIDS patients with those of isolates from potable water. *J. Clin. Microbiol.* 37: 1008-1012.

Barker, L.P., K.M. George. S. Falkow, and P.L. Small. 1997. Differential trafficking of live and dead *Mycobacterium marinum* organisms in macrophages. *Infect. Immun.* 65:1497-1504.

Beggs, M.L., R. Stevanova, and K.D. Eisenach. 2000. Species identification of *Mycobacterium avium* Complex isolates by a variety of molecular techniques. *J. Clin. Microbiol.* 38:508-512.

Biet, F., M.L. Boschiroli, M.F. Thorel, and L.A. Guilloteau. 2005. Zoonotic aspects of *Mycobacterium bovis* and *Mycobacterium avium-intracellulare* complex (MAC). *Vet. Res.* 36:411-436.

Bodle, E.E., J.A. Cunningham, P. Della-Latta, N.W. Schluger, and L. Saiman. 2008. Epidemiology of nontuberculous mycobacteria in patients without HIV infection, New Youk City. *Emerg. Infect. Dis.* 14:390-396.

Boulos, L., M. Prevost, B. Barbeau, J. Coallier, and R. Desjardins. 1999. LIVE/DEAD® BacLight™: application of a new rapid staining method for direct enumeration of viable and total bacteria in drinking water. *J. Microbiol. Methods* 37:77-86.

Chamberlin, W., D.Y. Graham, K. Hulten, H.M. El Zimaity, M.R. Schwartz, S. Naser, I. Shafran, and F.A. El Zaatari. 2001. Review article: *Mycobacterium avium* subsp. *paratuberculosis* as one cause of Crohn's disease. *Aliment. Pharmacol. Ther.* 15:337-346.

Cirillo, J.D., S. Falkow, L.S. Tompkins, and L.E. Bermudez. 1997. Interaction of *Mycobacterium avium* with environmental amoebae enhances virulence. *Infect. Immun.* 65:3759-3767.

Cruciani, M., C. Scarparo. M. Malena. O. Bosco, G. Serpelloni, and C. Mengoli. 2004. Meta-analysis of BACTEC MGIT 960 and BACTEC 460 TB, with or without solid media, for detection of mycobacteria. *J. Clin. Microbiol.* 42:2321-2325.

Dorshorst, N.C., M.T. Collins, and L.E. Lombard. 2006. Decision analysis model for paratuberculosis control in commercial dairy herds. *Prev. Vet. Med.* 75, 92-122.

Falkinham III, J.O. 1996. Epidemiology of infection by nontuberculous mycobacteria. *Clin. Microbiol. Rev.* 9:177-215.

Falkinham III, J.O. 2002. Nontuberculous mycobacteria in the environment. *Clinics in Chest Medicine* 23:529-551.

Falkinham III, J.O., C.D. Norton, and M.W. LeChevallier. 2001. Factors influencing numbers of *Mycobacterium avium, Mycobacterium intracellulare*, and other mycobacteria in drinking water distribution systems. *Appl. Env. Microbiol.* 67:1225-1231.

Feller, M., K. Huwiler, R. Stephan, E. Altpeter, A. Shang, H. Furrer, G.E. Pfyffer, T. Jemmi, A. Baumgartner, and M. Egger. 2007. *Mycobacterium avium* subspecies *paratuberculosis* and Crohn's disease: a systematic review and meta-analysis. *Lancet Infect. Dis.* 7:607-613.

Gagnon, G.A. and R.M. Slawson. 1999. An efficient biofilm removal method for bacterial cells exposed to drinking water. *J. Microbiol. Methods* 34:203-214.

Garrigo, M., L.M. Aragon. F. Alcaide, S. Borrell, E. Cardenosa, J.J. Galan, J. Gonzalez-Martin, N. Martin-Casabona, C. Moreno, M. Salvado, and P. Coll. 2007. Multicenter laboratory evaluation of the MB/BacT Mycobacterium Detection System and the BACTEC MGIT 960 System in comparison with the BACTEC 460TB System for susceptibility testing of *Mycobacterium tuberculosis*. *J. Clin. Microbiol.* 45:1766-1770.

Gasol, J.M., U.L. Zweifel, F. Peters, J.A. Fuhrman, and A. Hagstrom. 1999. Significance of size and nucleic acid content heterogeneity as measured by flow cytometry in natural planktonic bacteria. *Appl. Env. Microbiol.* 65:4475-4483.

Gerba, C.P., N. Nwachuku, R. Kelly. 2003. Disinfection resistance of waterborne pathogens on the United States Environmental Protection Agency's Contaminant Candidate List (CCL). *Journal of Water SRT - Aqua.* 52:81-94.

Glover, N., A. Holtzman, T. Aronson, S. Froman, O.G.W. Berlin, P. Dominguez, K.A. Kunkel, G. Overturf, G.J. Stelma, and C. Smith. 1994. The isolation and identification of *Mycobacterium avium* complex (MAC) recovered from Los Angeles potable water, a possible source of infection in AIDS patients. *Int. J. Environ. Health Res.* 4:63-72.

Grant, I.R. 2005. Zoonotic potential of *Mycobacterium avium* ssp. *paratuberculosis*: the current position. *J. Appl. Microbiol.* 98:1282-1293.

Greenstein, R.J. 2003. Is Crohn's disease caused by a *Mycobacterium*? Comparisons with leprosy, tuberculosis, and Johne's disease. *Lancet Infect. Dis.* 3:507-514.

Harris, N.B. and R.G. Barletta. 2001. *Mycobacterium avium* subsp. *paratuberculosis* in veterinary medicine. *Clin. Microbiol. Rev.* 14:489-512.

Hermon-Taylor, J., T.J. Bull, J.M. Sheridan, J. Cheng, M.L. Stellakis, and N. Sumar. 2000. Causation of Crohn's disease by *Mycobacterium avium* subspecies *paratuberculosis*. *Can. J. Gastroenterol.* 14:521-539.

Hett, E.C., and E.J. Rubin. 2008. Bacterial growth and cell division: a mycobacterial perspective. *Microbiol. Mol. Biol. Rev.* 72:126-156.

Hilborn, E.D., T.C. Covert. M.A. Yakrus, S.I. Harris, S.F. Donnelly, E.W. Rice, S. Toney, S.A. Bailey, and G.N. Stelma, Jr. 2006. Persistence of nontuberculous mycobacteria in a drinking water system after addition of filtration treatment. *Appl. Env. Microbiol.* 72: 5864-5869.

Hillemann, D., E. Richter, and S. Rusch-Gerdes. 2006. Use of the BACTEC Mycobacteria Growth Indicator Tube 960 Automated System for recovery of mycobacteria from 9,558 extrapulmonary specimens, including urine samples. *J. Clin. Microbiol.* 44:4014-4017.

Hoffman, R.M., C. Chauret, J.H. Standridge, and L. Peterson. 1999. Evaluation of four commercial antibodies. *J. Am. Water Works Assoc.* 91:69-78.

Huang, T.S., C.S. Chen, S.S.-J. Lee, W.K. Huang, and Y.C. Liu. 2001. Comparison of the BACTEC MGIT 960 and BACTEC 460TB Systems for detection of mycobacteria in clinical specimens. *Ann. Clin. Lab. Sci.* 31:279-283.

Inderlied, C.B., C.A. Kemper, and L.E.M. Bermudez. 1993. The *Mycobacterium avium* complex. *Clin. Microbiol. Rev.* 6:266-310.

Johansen, T.B., B. Djonne, M.R. Jensen, and I. Olsen. 2005. Distribution of IS*1311* and IS*1245* in *Mycobacterium avium* subspecies revisited. *J. Clin. Microbiol.* 43:2500-2502.

Kamala, T., C.N. Paramasivan, D. Herbert, P. Venkatesan, and R. Prabhakar. 1994. Evaluation of procedures for isolation of nontuberculous mycobacteria from soil and water. *Appl. Env. Microbiol.* 60:1021-1024.

Keller, A.P., M.L. Beggs, B. Amthor, F. Bruns, P. Meissner, and W.H. Haas. 2002. Evidence of the presence of IS*1245* and IS*1311* or closely related insertion elements in nontuberculous mycobacteria outside of the *Mycobacterium avium* Complex. *J. Clin. Microbiol.* 40:1869-1872.

Kirschner, R.A., B.C. Parker, and J.O. Falkinham III. 1999. Humic and fulvic acids stimulate the growth of *Mycobacterium avium*. *FEMS Microbiology Ecology* 30:327-332.

Kirschner, R.J., B.C. Parker, and J.I. Falkinham. 1992. Epidemiology of infection by nontuberculous mycobacteria. *Mycobacterium avium*, *Mycobacterium intracellulare*, and *Mycobacterium scrofulaceum* in acid, brown-water swamps of the southeastern United States and their association with environmental variables. *Am. Rev. Resp. Dis.* 145:271-275.

Kruuner, A., M.D. Yates, and F.A. Drobniewski. 2006. Evaluation of MGIT 960-based antimicrobial testing and determination of critical concentrations of first- and second-line antimicrobial drugs with drug-resistant clinical strains of *Mycobacterium tuberculosis*. *J. Clin. Microbiol.* 44:811-818.

Lambrecht, R.S., J. Carriere, and M.T. Collins. 1988. A model for analyzing growth kinetics of a slowly growing *Mycobacterium* sp. *Appl. Env. Microbiol.* 54:910-916.

Lambrecht, R.S., S. Schroederus, H. Yang, and M.T. Collins. 1995. The need for employing standardized susceptibility testing methods and interpretive criteria which provide optimal information for evaluating the response of intestinal mycobacterial isolates to anti-infective agents. In: Chiodini, R.J., Collins, M.T., Bassey, E.O.E. (Eds), Proceedings of the Fourth International Colloquium on Paratuberculosis. International Association for Paratuberculosis, Rehoboth, MA, p. 149-(Abstract).

Langsrud, S. and G. Sundheim. 1996. Flow cytometry for rapid assessment of viability after exposure to quaternary ammonium compound. *J. Appl. Bacteriol.* 81:411-418.

Le Dantec, C., J.P. Duguet, A. Montiel, N. Dumoutier, S. Dubrou, and V. Vincent. 2002. Occurrence of mycobacteria in water treatment lines and in water distribution systems. *Appl. Env. Microbiol.* 68:5318-5325.

Leclerc, H., L. Schwartzbrod, and E. Dei-Cas. 2002. Microbial agents associated with waterborne diseases. *Critical Rev. Microbiol.* 28: 371-409.

Lehtola, M.J., E. Torvinen, J. Kusnetsov, T. Pitkanen, L. Maunula, C.H. von Bonsdorff, P.J. Martikainen, S.A. Wilks, C.W. Keevil, and I.T. Miettinen. 2007. Survival of *Mycobacterium avium*, *Legionella pneumophila*, *Escherichia coli*, and caliciviruses in drinking water-associated biofilms grown under high-shear turbulent flow. *Appl. Env. Microbiol.* 73:2854-2859.

Neumann, M., R. Schulze-Roebbecke, C. Hagenau, and K. Behringer. 1997. Comparison of methods for isolation of mycobacteria from water. *Appl. Env. Microbiol.* 63:547-552.

Norton, C.D., M.W. LeChevallier, J.O. Falkinham III. 2004. Survival of *Mycobacterium avium* in a model distribution system. *Water Res.* 38:1457-1466.

Primm, T.P., C.A. Lucero, and J.O. Falkinham III. 2004. Health impacts of environmental mycobacteria. *Clin. Microbiol. Rev.* 17:98-106.

Promega. BacTiter-Glo® Microbial Cell Viability Assay. Technical Bulletin. 9/07. 2008. Madison, WI, Promega Corp. 1-9-2007.

Rastogi, N., E. Legrand, and C. Sola. 2001. The mycobacteria: an introduction to nomenclature and pathogenesis. *Revue Scientifique et Technique* 20:21-54.

Rastogi, N., M.-C. Potar, and H.L. David. 1988. Action of antituberculous and B-lactam drugs (including imipenem) against extra- and intra-cellularly growing *Mycobacterium avium-intracellulare. Ann. Inst. Pasteur/Microbiol.* 139:225-232.

Reddacliff, L.A., P.J. Nicholls, A. Vadali, and R.J. Whittington. 2003. Use of growth indices from radiometric culture for quantification of sheep strains of *Mycobacterium avium* subsp. *paratuberculosis*. *Appl. Env. Microbiol.* 69:3510-3516.

Richter, E., J. Wessling, N. Lugering, W. Domschke, and S. Rusch-Gerdes. 2002. *Mycobacterium avium* subsp. *paratuberculosis* infection in a patient with HIV, Germany. *Emerg. Infect. Dis.* 8:729-731.

Rubery, E. 2001. A review of evidence for a link between exposure to *Mycobacterium paratuberculosis* (MAP) and Crohn's disease (CD) in humans. Food Standards Agency, UK (55 pages).

Schulze-Roebbecke, R., A. Weber, and R. Fischeder. 1991. Comparison of decontamination methods for the isolation of mycobacteria from drinking water samples. *J. Microbiol. Methods* 14: 177-183.

Shin, S.J., and M.T. Collins. 2008. Thiopurine drugs azathioprine and 6-mercaptopurine inhibit *Mycobacterium paratuberculosis* growth in vitro. *Antimicrob. Agents Chemother.* 52:418-426.

Shin, S.J., J.H. Han, J.H., E.J.B. Manning, and M.T. Collins. 2007. Rapid and reliable method for quantification of *Mycobacterium paratuberculosis* by use of the BACTEC MGIT 960 system. *J. Clin. Microbiol.* 45:1941-1948.

Somoskovi, A., C. Kodmon, A. Lantos, Z. Bartfai, L. Tamasi, J. Fuzy, and P. Magyar. 2000. Comparison of recoveries of *Mycobacterium tuberculosis* using the automated BACTEC MGIT 960 System, the BACTEC 460 TB System, and Lowenstein-Jensen medium. *J. Clin. Microbiol.* 38:2395-2397.

Stabel, J.R., 1997. An improved method for cultivation of *Mycobacterium paratuberculosis* from bovine fecal samples and comparison to three other methods. *J. Vet. Diagn. Invest.* 9:375-380.

Steinert, M., K. Birkness, E. White, B. Fields, F. Quinn. 1998. *Mycobacterium avium* bacilli grow saprozoically in coculture with *Acanthamoeba polyphaga* and survive within cyst walls. *Appl. Env. Microbiol.* 64:2256-2261.

Sung, N., and M.T. Collins. 1998. Thermal tolerance of *Mycobacterium paratuberculosis*. *Appl. Env. Microbiol.* 64:999-1005.

Taylor, R.H., J.O. Falkinham III, C.D. Norton, and M.W. LeChevallier. 2000. Chlorine, chloramine, chlorine dioxide, and ozone susceptibility of *Mycobacterium avium*. *Appl. Env. Microbiol.* 66:1702-1705.

Turenne, C.Y., D.M. Collins, D.C. Alexander, and M.A. Behr. 2008. *Mycobacterium avium* subsp. *paratuberculosis* and *M. avium* subsp. *avium* are independently evolved pathogenic clones of a much broader group of *M. avium* organisms. *J. Bacteriol.* 190:2479-2487.

Turenne, C.Y., R. Wallace Jr., and M.A. Behr. 2007. *Mycobacterium avium* in the postgenomic era. *Clin. Microbiol. Rev.* 20:205-229.

U.S. Department of Agriculture, 1974. Laboratory Methods in Veterinary Mycobacteriology. Mycobacteriology Unit, National Veterinary Services Laboratories. APHIS, USDA, Ames, Iowa.

von Reyn, C.F., R.D. Arbeit, C.R. Horsburgh, M.A. Ristola, R.D. Waddell, S.M. Tvaroha, M. Samore, L.R. Hirschhorn, J. Lumio, A.D. Lein, M.R. Grove, and A.N. Tosteson. 2002. Sources of disseminated *Mycobacterium avium* infection in AIDS. *J. Infect.* 44:166-170.

von Reyn, C.F., J.N. Maslow, T.W. Barber, J.O. Falkinham III, and R.D. Arbeit. 1994. Persistent colonization of potable water as a source of *Mycobacterium avium* infection in AIDS. Lancet 343:1137-1141.

Wayne, L.G., and L.G. Hayes. 1996. An in vitro model for sequential study of shiftdown of *Mycobacterium tuberculosis* through two stages of nonreplicating persistence. *Infect. Immun.* 64:2062-2069.

Whan, L., I.R. Grant, and M.T. Rowe. 2006. Interaction between *Mycobacterium avium* subsp. *paratuberculosis* and environmental protozoa. *BMC Microbiol* 6:63-68.

Whan, L.B., I.R. Grant, H.J. Ball, R. Scott, and M.T. Rowe. 2001. Bactericidal effect of chlorine on *Mycobacterium paratuberculosis* in drinking water. *Letters Appl. Microbiol.* 33:227-231.

Whipple, D.L. and R.S. Merkal. 1983. Modifications in the technique for cultivation of *Mycobacterium paratuberculosis.* In: Merkal, R.S. (Ed.), Proceedings of the First International Colloquium on Research in Paratuberculosis. National Animal Disease Center, Ames, Iowa, pp. 80-90.

Whittington, R.J., I.B. Marsh, P.J. Taylor, D.J. Marshall, C. Taragel, and L.A. Reddacliff. 2003. Isolation of *Mycobacterium avium* subsp *paratuberculosis* from environmental samples collected from farms before and after destocking sheep with paratuberculosis. *Aust. Vet. J.* 81:559-563.

Whittington, R.J., D.J. Marshall, P.J. Nicholls, I.B. Marsh, and L.A. Reddacliff. 2004. Survival and dormancy of *Mycobacterium avium* subsp. *paratuberculosis* in the environment. *Appl. Env. Microbiol.* 70:2989-3004.

ABBREVIATIONS

APHA	American Public Health Association
AWWA	American Water Works Association
AwwaRF	Awwa Research Foundation
BAR	Biofilm annular reactor
CFU	colony forming unit
CGI	cumulative growth unit
cm	centimeter
CSCC	continuous separation channel centrifugation
°C	degrees Celsius
CCT	contaminant candidate list
CSU	Colorado State University
CT	contact time concentration of chlorine
Ct	cycle time (PCR)
FITC	fluorescein isothiocyanate
HDPE	high-density polyethylene
IMS	immunomagnetic separation
IS	insertion sequence
JTC	Johne's Testing Center, University of Wisconsin-Madison
L	liter
MAA	*Mycobacterium avium* ss. *avium*
MAC	*Mycobacterium avium* Complex
MAP	*Mycobacterium avium* ss. *paratuberculosis*
MF	mean fluoresence (flow cyotometry)
mg	milligram
mg/L	milligram per liter
MGIT	mycobacteria detection system (Becton-Dickinson)
min	minute
mL	milliliter
µL	microliter
mL/min	milliliter per minute
mm	millimeter
µm	micrometer
nm	nanometer
N	concentration of pathogens in treated water

PAC	Project Advisory Committee
PBS	phosphate buffer solution
PCTE	polycarbonate trac ketch (filter)
sec	second
Tris	trishydroxymethylaminomethane (buffer solution)
TOC	total organic carbon
UW	University of Wisconsin-Madison
U.S.	United States
USEPA	United States Environmental Protection Agency
UK	United Kingdom
WI	Wisconsin
WSLH	Wisconsin State Laboratory of Hygiene

IWA Publishing's authorised EU representative for General Product Safety Regulations is Diane D'Arras, 15 rue Duret, 75116 Paris, France, e-mail: safety@iwap.co.uk.

Printed and bound by CPI Group (UK) Ltd, Croydon, CR0 4YY
17/09/2026
02218813-0010